Is Progress Speeding Up?

Is Progress Speeding Up?

Our Multiplying Multitudes of Blessings

John M. Templeton

Templeton Foundation Press

Philadelphia & London

TEMPLETON FOUNDATION PRESS
Two Radnor Corporate Center, Suite 330
100 Matsonford Road
Radnor, Pennsylvania 19087

Printed in the United States of America

Library of Congress 97-090756

ISBN 1-890151-02-5

Acknowledgments can be found on p. 271.

Contents

	Introduction	1
1	How We Live	15
2	Food	31
3	Health and Life Expectancy	49
4	Working Conditions	69
5	Technology	87
6	Political Freedom	111
7	Economic Freedom	127
8	Education	147
9	Information and Communications	167
10	Transportation	185
11	Leisure	193
12	The Environment	211
13	Getting Along	229
14	The Spirit	243
15	The Future	259
	Bibliography	267
	Acknowledgments	271
	Index	275

Is Progress Speeding Up?

Introduction

"Our species is better off in just about every measurable material way. And there is stronger reason than ever to believe that these progressive trends will continue past the year 2000, past the year 2100, and indefinitely."
—Julian L. Simon, *The State of Humanity*, 1995

- U.S. employment increased by 18 million jobs in the last decade.
- Life expectancy is at a record high around the world.
- America is witnessing the creation of history's first mass upper class.
- More than 50 million Americans now own stock in individual companies or shares in stock mutual funds.

A Wonderful Time in History

I was born to a family of limited means in a small Tennessee town on November 29, 1912, which was a Friday and a very long time ago. Rumblings that would lead to World War I had begun to be heard over the horizon. In the United States, Woodrow Wilson had just won the presidential election by a landslide. In Arizona, Wisconsin, and Kansas, women had gained the right to vote. A loaf of bread cost only a nickel, a gallon of milk 35¢, a new car $500, and a new home $4,800.

I need hardly point out that much has changed in the world in the more than eight decades since my birth. And on balance, I firmly believe that things have changed for the better. Indeed, I am convinced that people—not just in the United States, but around the world—are vastly better off today than ever before, and that this welcome trend will continue.

This is a wonderful time to be alive! And this book will document that central fact from many vantage points—in terms of our nutrition and health, our living standards and working conditions, our political and economic freedoms, our educational facilities and our ability to communicate with one another, our ease of movement and our leisure, and, most importantly, our ability to get along with one another and with our Creator.

This book, in sum, is a documentation—and a celebration—of the progress of humankind during my lifetime. I hope that you the reader will find it highly informative, and also that it will serve as a strong antidote to the poisonous pessimism that invades so much of what we hear and read about the time in which we live.

It is distressing that the media, and thus much of the public, dwell so heavily on bad news. If one were to judge the state of humanity solely by the daily headlines and news broadcasts, one would come away convinced that we inhabit a world besieged by poverty, insecurity, conflict, and oppression. There is no denying that ills exist within modern society, especially in the so-called Third World of poorer nations. However, with the attention that the media gives them, these ills are masking the stupendous progress being made in so many areas.

To wit, people today, on the average, are better fed, better clothed, better housed, and better educated than at any previous time. Fewer and fewer people live under the weight of tyranny. In most parts of the world, people are enjoying longer, healthier, more fulfilling lives. And yet, unfortunately, these very real triumphs generally pass unnoticed, which is a very great pity. Why must so many people be so unhappy when in reality humankind is living in the most glorious period in all of history?

This unwarranted pessimism encourages cynicism, discourages effort, particularly among the young, and disrupts community relationships. If people look for the bad in the world around them—in their leaders, in their neighbors, in their personal situations—they will surely find the bad, and this can have destructive consequences. Dreams are not built on cynicism; optimism begets achievement.

I am convinced that if people look for the good, they will find it, and that in countless ways this will have *constructive* consequences. Enthusiasm breeds effort and success. This book is a small attempt to encourage people to look for the good, for it is there in abundance to be seen and appreciated.

Today's problems are often viewed without perspective. Much is made, for example, of the persistence of poverty in the world, but gloom-seers forget that "poverty" is a relative term. In the United States, the definition of poverty has changed with growing prosperity; most people classified as "poor" today are materially better off than middle-class and even some wealthy Americans were several decades ago.

Even though the poverty "threshold" in the United States is set higher today than years ago, fewer and fewer Americans are classified as living in poverty, which is defined as three times food expenses, as calculated by the Department of Agriculture. In 1995, the latest year for which statistics are available, the poverty level for a family of four came to nearly $16,000, not including government benefits such as Medicare. By this standard, 13.8 percent of the U.S. population was below the poverty line in 1995, down from 14.5 percent a year earlier and 15 percent in 1993. At the same time, people generally enjoy unprecedented access to education, medical services, transportation, food, and housing—access that most families would have envied when I was born.

In the case of food, the world's population of nearly six billion people is, on the whole, far better fed than at any previous time. Never before has food been so plentiful, so nutritious, or so available in such variety. All this is made possible by highly sophisticated agricultural equipment, large-scale commercial farming, new cultivation techniques, improved fertilizers and pesticides, and genetically engineered plants and even animals.

In the last century, families in Europe spent more than 75 percent of their earnings on groceries alone, while at present they spend less than 10 percent. Additionally, the world's population now consumes about 20 percent more food per person than a half century ago. Only twenty-five years ago, people in the poorer developing countries were able to obtain only about four-fifths of their minimum daily dietary needs, on the average, whereas today they receive substantially more than the daily minimum.

Evidence of progress is widespread. In the field of health, river blindness, a cause of immense suffering and social disruption in much of Africa for centuries, has been virtually eliminated during the past ten years.

Likewise, current worrying over the pollution of our environment ignores extensive evidence of improved environmental quality in the world's most industrialized nations. London's "killer smogs," for instance, are no longer a threat to residents of that wonderful city. London buildings that were black with soot when I was an Oxford student now are white and colorful.

Complaints about airport delays and traffic-clogged highways miss the central point that travel today is vastly swifter and more comfortable than in earlier decades.

In today's schools, teenagers easily comprehend matters in science that would have confounded their parents and their parents' teachers.

People today are said to be less caring about each other. But if that is so, how does one explain the fact that charitable giving leaps from new record to new record? Americans nowadays contribute some $60 billion yearly to religious groups alone, which is roughly fifteen times more than they spend on professional baseball, football, and basketball *com-*

bined. Moreover, yearly attendance at religious functions to-
tals nearly six billion, dwarfing the combined total for baseball, football, and basketball.

At the same time, wealthier nations are digging deeper
into their pockets to help poorer nations. For instance, officials of the so-called Group of Seven, comprising the United
States and six other major industrial nations, recently approved a $55 billion fund to help less prosperous neighbors
cope with financial difficulties.

Encouraging Long-Term Trends

Standing back from daily headlines helps us focus on many
heartening developments that belie the pessimistic percep-
tions that are so pervasive. Work force reductions, or "downsizing," within some of the largest corporations, for example,
have engendered a widespread sense of economic insecurity
among Americans. However, the process constitutes a necessary, healthy, normal adjustment to global competition and
new technologies.

In fact, corporate downsizing is neither a new nor a wors-
ening problem. Some 3.8 million American workers were
displaced in 1993–95 from jobs that they had held at least
three years. But this was 718,000 *fewer* displacements than
occurred in 1991–93. Moreover, some 70 percent of the displaced workers were reemployed within three months, and in
about half of these cases the new jobs paid as much as or
more than the old jobs.[1]

"We're constantly fretting about the 'victims' of downsizing [but] we seldom celebrate the beneficiaries of upsizing,"
says Peter Lynch, the highly successful mutual fund man-

---

1. Department of Labor, Bureau of Labor Statistics, "Worker Displacement
During the Mid-1990s," USDL report 96-336, August 22, 1996.

Leisure time is increasing along with pay levels, bringing countless benefits to the quality of our lives. It grants new opportunities for pleasure and learning to millions of people whose forebears knew little else away from home but on-the-job drudgery. Moreover, increasing leisure serves to strengthen overall economic activity. Americans now spend nearly $3 billion a year on exercise equipment, some $10 billion on athletic clothing of various sorts, $3.7 billion on bicycles, and $1.4 billion on golf equipment. Nearly 60 million Americans fish in fresh or salt water, 23 million play golf, 15 million ski downhill or cross country, 41 million bowl, 48 million ride bicycles, and 50 million backpack or camp out.

Meanwhile, working conditions around the globe, on the whole, are safer and more pleasant than ever before. No longer is the typical worker a meaningless cog in a mindless corporate machine. In the modern workplace workers share a real voice in how their businesses should operate. This adds to the challenge of jobs and makes them more interesting.

In the chapters that follow you will also observe, through broad-based data as well as in telling anecdotes, the remarkable spread of political and economic freedoms in countries less fortunate in recent decades than, say, the United States. Today, more people live in democratic societies than ever before. Indeed, this decade is the first in which democratic nations are in the majority. Democracy has swept across Eastern Europe, the former Soviet Union, and much of Latin America, while military regimes in more than a dozen African nations have stepped aside for free elections.

At the same time, economic freedom is blossoming in areas where individual enterprise and private business activity were once severely constrained. Inefficient state industries are being privatized, creating new investment opportunities. Oppressive barriers to trade between nations are being dismantled. As a result, the volume of world exports has grown more than eleven-fold in the past four decades.

For individuals, today's investment opportunities are unprecedented, as you will see in Chapter Seven. The world's great financial and foreign-exchange markets have long served wealthy investors wishing to buy and sell securities or currencies. But now, through an expanding variety of financial intermediaries, these same markets also increasingly serve investors of limited means.

In all, there are some 10 million private businesses in the United States alone, more than four times the number as recently as 1970. New businesses are springing up at the remarkable rate of some 700,000 yearly. So there is no shortage of businesses in which to invest for the more than 50 million Americans who now own stock in individual companies or shares in stock mutual funds.

This is also a great time to be alive for reasons that transcend our material well-being. The Cold War has ended. The former Soviet Union has broken apart. The Berlin Wall is demolished and Germany is reunited. Seemingly intransigent foes are progressing toward peace in the Middle East. The United Nations provides a forum in which nations can seek peaceful solutions to disputes that once might have led to armed conflict. As never before, the nuclear superpowers strive to avoid confrontations that could bring on war—between smaller nations as well as one another. Increasingly, people around the world perceive that war is no solution to international disagreements.

In brief, people are getting along better than ever before. Laws in the United States, for example, now generally ensure equal opportunity in employment, as well as freedom from racial discrimination and sexual harassment. With giving at an all-time high, charitable foundations help disadvantaged members of society in countless ways, from ensuring their civil rights to helping them prepare for the work place.

Meanwhile, religion flourishes around the world, reviving most notably within the once-Communist nations of the former Soviet bloc. In the United States, the number of inspirational books published annually has grown one hundredfold in the course of this century. Moreover, as many as forty organizations now publish newsletters exploring various connections between religious belief and science. A recent poll found that four of every five Americans believe spiritual faith can help people recover from illness, and nearly six in every ten report that their religious faith has helped them at one time or another recover from an illness.[4]

Other recent surveys turn up an unmistakable correlation between religious faith and physical well-being. One study shows, for example, a connection between religious commitment and lower blood pressure and another suggests a linkage between church-going and low rates of mental disorder. Accordingly, the prestigious Harvard Medical School recently launched a symposium to study possible ties between spirituality and healing.

At the same time, dozens of books are appearing that emphasize the importance of ethics in business affairs and explore linkages between religious teachings and how the marketplace works. One such book, *The Seven Spiritual Laws of Success,* recently ranked as the top-selling business book in the United States. Other recent business/religious books that have sold well include such titles as *The Management Methods of Jesus* and *Jesus C.E.O.*[5]

Pessimism: Unwarranted but Stubborn

Why, I keep asking myself, are so many people so pessimistic about the time in which we live when in reality we are so blessed in so many ways?

4. Tom McNichol, "The New Faith in Medicine," *USA Weekend*, April 5–7, 1996.
5. Deborah Stead, "Mixing Religion and Capitalism," *New York Times*, April 7, 1996.

There are, I believe, several answers to this troubling question. One reason, simply stated by a veteran journalist friend of mine, is that *bad news sells*. Readers of newspapers and magazines, as well as TV viewers and radio listeners, are more apt to show an interest if the news is negative. Dagwood, in the *Blondie* strip below, is clearly a rare exception to this rule.

There is nothing particularly new about this very human tendency to focus on bad news. It existed long before TV sets were ubiquitous in the home or portable radios brought news to people far from any newsstand. What has changed, however, is that today the opportunity to read or see or hear the news is unprecedented. Around major cities, men and women commuting to and from work in their automobiles routinely get the latest news on their car radios. In the remote countryside, agricultural workers tune in to the news using handheld radios. When I was young, on the other hand, there was no television to watch or radio to hear, and so we weren't subjected to the same barrage of negative reports that engulfs us today.

Interestingly, the general public appears well aware that the news is steeply tilted toward pessimism. A recent Louis Harris & Associates poll found that 90 percent of the U.S. population agree that the media are likelier to report "about terrible, violent crimes" than the "good news" that violent crime is decreasing. The poll also found that 75 percent of the popu-

Reprinted with special permission of King Features Syndicate.

lation agree that the media prefer to focus on the news that "people are losing their jobs" instead of on the "good news" that many more jobs are being created than eliminated.[6]

Perhaps the clearest evidence of pessimism's selling power occurs in the field of book publishing. Books predicting all sorts of gloom and doom perennially roll off the presses and sell well. Recent examples include such titles as *Bankruptcy 1995* and *The Great Depression of 1990*. For all of the nonsense that they put forward—America did not go bust and there was no depression in 1990—both books reached the top of various best-seller lists.

Meanwhile, such books as Julian Simon's decidedly upbeat *The State of Humanity*, which I quoted at the start of this chapter, languish in obscurity even though, as Professor Simon points out in his introduction, "the years have been kind to our [optimistic] forecasts—or more importantly, the years have been good for humanity."[7] In fact, Professor Simon is so confident about his optimism that he and some fellow optimists have assembled $100,000 to wager with pessimists, with a limit of $10,000 for any single bet. Pick any year after the turn of the century and they will bet you that the average person in any nation will live longer, eat better, be better educated, have a bigger home, and enjoy a safer, cleaner environment than at present.[8] Professor Simon, I must warn you, has won a number of similar bets in the past.

As an optimist myself, I should add that it is even possible to find a possible long-term benefit in the media's excessive attention to bad news. After all, by focusing on troubling trends, we may be likelier to take whatever actions may be necessary to correct them.

6. Richard Morin, "The Blame Game," *Washington Post Weekly Edition*, September 16–22, 1996.

7. Julian L. Simon, *The State of Humanity,* (Blackwell Publishers Inc., Cambridge, Mass., 1995).

8. John Tierney, "The Optimists Are Right," *New York Times Magazine*, September 29, 1996.

The highly optimistic book that you hold in your hands at this moment, I should report, was offered to more than a dozen prominent publishing houses. All but one turned it down flat, without even a suggestion that the book might be publishable if I were to rework it. The lone exception, a large, influential publishing house, indicated that it might consider publishing the book, but only if I would introduce a great deal of highly negative material—an offer that I promptly declined.

Well, the book that you hold obviously has found a publisher. If, after reading it, you come away with a greater sense of optimism and a more constructive view about today's world, I will be most grateful.

1.

How We Live

"There's place and means for every man alive."
—William Shakespeare (1564–1616), *All's Well That Ends Well*

- One million American families have incomes exceeding $200,000.
- Chinese entrepreneurs establish new businesses every eleven minutes in Shanghai alone.
- Real per capita GDP in the United States stands at a record level.

The Huge Rise in Living Standards

A great deal of subjectivity comes into play when people start talking about living standards. To understand why, let's consider the hypothetical case of someone—we'll call him John Jones—whose living standard, by all appearances, is decidedly on the rise.

A little while ago, John received a very big pay rise. As a result, he now drives a more powerful, roomier car than the one he traded in. Moreover, he recently moved into a bigger house in a better neighborhood. John also manages nowadays to take his wife out to dinner a couple of times a week, while a year ago such treats were limited to about once a month. And he now can afford to eat costlier food on these outings, such as filet mignon instead of pasta. In addition, this year's family vacation will be at Montego Bay in beautiful Jamaica, whereas last year's trip was to Montauk on the tip of Long Island. And, most importantly, his two children have switched from public to private schools.

By most conventional yardsticks, John's living standard has risen—bigger car, better house, more pay, fancier meals, a trip to the Caribbean, private schools for the kids. Can there be any doubt that John and his family are better off?

It is possible to question, for instance, the benefits of having a bigger, more powerful car. After all, it soaks up more gasoline and costs more to insure and maintain. And who needs more red meat? John's cholesterol count already is on the high side and it would be better for him to stick with pasta. As for vacations, the food and streets are safer in Montauk than Jamaica. As for schools, public ones may prepare youngsters better for getting along—and ahead—in these egalitarian times.

That said, it is possible, by combining statistical results with anecdotal evidence, to gain a pretty clear sense at least of the *direction* in which living standards have been moving in the United States, as well as in most other countries in recent decades. This direction has been up and, for all the talk to the contrary in the press and elsewhere, the rise has been remarkably sharp and is continuing.

Let's consider the broadest economic measure—the gross domestic product (GDP)—in per capita terms to allow for population growth and in so-called real terms to allow for increases due merely to rising prices. Real per capita GDP in the United States now stands at a record level—slightly above $20,000, in terms of the dollar's 1987 buying power. This is up from only $6,857 in 1940 and $9,352 in 1950. Thus, using GDP as a yardstick, the U.S. standard of living has more than doubled since mid-century and roughly tripled since before the U.S. entry into World War II. The bulk of this increase reflects all manner of goods and services that enhance people's living standards, from better-quality automobiles to safer surgery.

There is much talk of a stagnation, even a decline, in living standards since the mid-1970s. But GDP data show nothing of the sort. In 1975, for instance, real per capita GDP amounted to $14,917, again in terms of the dollar's 1987 buying power. The rise since then approaches 40 percent. A gain of such magnitude hardly spells stagnation!

While such things as traffic jams may falsely inflate the GDP numbers by adding to fuel consumption and vehicle wear and tear, a far more serious distortion tending to *deflate* the GDP numbers is the omission of work performed by unpaid workers, such as charity volunteers and spouses who keep house.

Other broad gauges tell a similar story. Since 1975, real per capita disposable income—what's left to spend or save after all taxes have been paid—has increased even more sharply than real per capita GDP. Sharper still has been the rise in real per capita consumer spending. These barometers, which are about as close as economists have thus far come to gauging living standards in a nation, denote a far happier situation than most news reports would have us believe is the case.

The most frequent and most publicized lament of living-standard doomsayers is that the median income of American households, adjusted for inflation, is lower today than in the mid-1970s. But in light of the per capita gains noted above, how can this be? In part, the explanation is that—for reasons ranging from later marriages to longer life expectancy—there are fewer people in the average American household today than, say, a quarter-century ago. The upshot is a reduction in average household income that reflects a shrinkage in household size, rather than a drop in living standards.

Other data also point to striking gains in living standards in America. In 1950, more than a third of U.S. dwellings lacked indoor plumbing and less than one in ten had a television set. As recently as 1960, roughly one home in every four still had no telephone. Today, less than one home in twenty has no telephone, only one home in one hundred has no plumbing, and only one in fifty has no TV set.

With such gains in living conditions, it is no surprise that American workers require far less time nowadays to earn what is needed to purchase various goods and services. The table below provides some examples of this pattern.

To be sure, not every item requires less work time now than years ago. The fee for an appendectomy amounts to about nine days of work, up from seven two decades ago, and a year's tuition at Brown University requires more than eight months of work, up from just over four months. Such examples, however, are exceptions. I should also point out, of course, that with new antibiotics appendectomies are now

Work Time Needed to Earn Enough to Buy Items (in minutes, unless noted)

	1974	1994
First class stamp	1.2	1.1
Gallon unleaded gasoline	6.2	4.8
Pound of chicken	7.1	5.6
Movie ticket	22.2	15.8
5-min. call, Dallas–Seattle	35.6	4.7
Big Mac, medium soda, & fries	13.5	10.3
Monopoly board game	46.7	37.7
Boston–New York train trip	141	128
Man's dress shirt from Sears	118	91
Hewlett-Packard calculator	3,823	265
Zenith 19-inch color TV set	3 weeks	3 days
American Airlines flight N.Y.–L.A. discount fare	4.25 days	3.1 days

SOURCE: *Wall Street Journal,* March 29, 1995

safer and rarer. By the same token, the quality of an education at a leading university such as Brown has clearly improved over the years.

Productivity's Role

In much of the post-World War II era, as we have seen, incomes have risen steeply even after inflation and population growth are taken into account, and this rise has led to great strides in living standards. Such progress would have been impossible, however, without solid gains in productivity, or the hourly output of the people turning out goods and services. Advances in productivity serve to offset the potential inflationary impact of pay increases. Without rising productivity, pay gains inevitably lead to higher labor costs, which force companies to raise prices and feed inflation. This, in turn, erodes any gains in paycheck buying power.

To see how productivity advances work to offset the inflationary impact of higher pay, and thus improve living standards, consider a hypothetical example—the case of Tina O'Toole, who works in a factory that turns out computer chips. Let's suppose that Tina receives a 10 percent raise in her hourly pay and then imagine three possible scenarios.

In scenario one, Tina's hourly production of chips climbs by the same amount as her pay. As a result, her pay boost will not alter the hourly cost of her labor. She will be paid 10 percent more each hour, but she also will be producing 10 percent more chips per hour. As a result, her company will be under no labor-cost pressure to raise prices; nor will it be tempted to lower them. If prices don't change, Tina's pay raise of 10 percent will serve to increase the buying power of her paycheck by a like amount, and so her living standard will improve.

In scenario two, Tina's hourly output of chips rises 15 percent, or more than the boost in her hourly pay. As a result, the per-chip cost of her labor will drop, despite her pay raise. This decline, in turn, will enable Tina's company to lower prices without hurting its profitability. If prices decline, Tina's buying power, and ultimately her living standard, will improve by an even greater margin than in scenario one—by 15 percent instead of 10 percent.

In scenario three, Tina's hourly output of chips stays the same after her raise. As a result, the cost of her labor will rise. Her employer will be paying her more each hour to produce the same number of chips that she turned out previously, when she was paid a lower hourly wage. This will put pressure on Tina's company to raise prices, which will spur inflation and wipe out the increase in buying power that Tina expected her raise to bring.

Happily, productivity in the United States, as well as most other major industrial nations, has advanced greatly over the years. The average output per hour of U.S. workers has approximately doubled since 1960. In the same period, by no coincidence, real hourly pay has climbed about 60 percent. Without the sharp advance in productivity, this gain in real pay would have been wiped out by inflation and living standards would have declined, rather than increased markedly.

Nowhere have the benefits of advancing productivity proved greater than on the farm. Today, less than 3 percent of the U.S. labor force works in agriculture, down from more than 20 percent as recently as the 1930s. Yet, today's food production easily meets demand, as improvements in agricultural machinery, crop varieties, pesticides, and fertilizers have helped triple the output of American farms in the past three decades. An added benefit of this productivity surge is that it has averted the clearing of vast forest areas—acreage that otherwise would have been needed to produce sufficient food for an expanding population.

Comfort and Wealth

The improvement in living standards in the United States is particularly apparent in housing data. Not only are homes equipped with conveniences unimagined years ago, from computers to VCRs, but the homes themselves are vastly improved. In 1970, one of every three single-family houses in the United States occupied less than 1,200 square feet and only one house in ten took up more than 2,400 square feet. Today, the numbers are nearly reversed. About one house in ten occupies less than 1,200 square feet and about one in three occupies at least 2,400 square feet. The median house size has increased in the period from less than 1,400 square feet to nearly 2,000.

Other improvements in housing abound. In 1970, only one-third of all single-family houses had fireplaces and only one-third contained central air conditioning. Today, two-thirds of all houses have fireplaces and four-fifths contain central air. Meanwhile, the percentage of houses equipped with garages has risen from less than 60 percent in 1970 to about 85 percent now. In sum, no people in history have been as well housed as Americans are today.

The present level of home ownership in the United States is unprecedented, I should add, despite much well-publicized concern about homeless individuals wandering the streets of New York, Los Angeles, and other large American cities. The number of occupied housing units, fast approaching 100 million, has increased nearly 20 percent just since 1980. This works out to nearly one home for every three people, and about two-thirds of these homes are lived in by their owners.

In all, today's American families typically occupy housing with amenities for which, I have no doubt, King Louis XIV would have gladly traded several wings of Versailles—indoor toilets, running water, telephones, electric and gas stoves and refrigerators, as well as central heating and air conditioning.

Additionally, swimming pools, decks, barbecues, and saunas have become commonplace in suburban America. These comfortable homes, moreover, represent a major investment whose value can help provide elderly Americans with a generous retirement-income supplement through such arrangements as reverse mortgages.

Indeed, housing is a major factor in the remarkable increase in the nation's overall wealth. This wealth comes in two forms: tangible items that we can touch, like houses or automobiles; and financial items, like shares of stocks or bonds. The table below tracks the growth in both tangible and financial wealth over the post-World War II era. The numbers are in terms of the buying power of the 1987 dollar, to allow for inflation.

The increase in total wealth, as the table indicates, has persisted throughout the postwar era, even through years

Growth in Wealth 1946–1993

	Total Household Wealth*	Tangible	Financial
1946	4,534.7	1,318.0	3,216.8
1950	4,817.5	1,768.2	3,049.3
1955	6,095.3	2,394.4	3,700.9
1960	6,974.7	2,796.9	4,177.8
1965	8,516.6	3,180.6	5,336.0
1970	9,275.9	3,778.1	5,497.8
1975	10,037.3	4,697.6	5,339.7
1980	12,802.8	6,229.1	6,573.6
1985	14,549.5	6,892.7	7,656.8
1990	16,472.6	7,584.0	8,888.6
1993	18,501.2	8,001.7	10,499.5

SOURCE: Federal Reserve System and U.S. Department of Commerce
*Figures are in billions of 1987 dollars

when financial wealth has fallen. In fact, the increase in the past two decades has been substantially sharper than in the previous two decades—an acceleration that belies the widespread notion that America has become stuck in a sort of economic quagmire.

Even though wealth has grown greatly in the United States, there is much concern that its distribution has become increasingly uneven, with the rich getting a great deal richer, while others are being left behind—if true, an unhealthy situation breeding social unrest. But is it true? The supposition is fed by reports in newspapers and magazines and on TV that detail the mounting assets and luxurious lifestyles of the very rich. In its "400" issue in 1995, *Forbes* magazine listed, among the four hundred wealthiest Americans, four whose net worth exceeded $5 billion. The cover of another *Forbes* issue (May 22, 1995) portrayed an overweight man reclining and stuffing himself with grapes in the manner of a Roman emperor. Above this scene ran the headline: "Pigging It Up: 800 Chief Executive Paychecks—Median Take $1.3 Million." Similarly, the Sunday magazine of the *New York Times* (November 19, 1995) documented the rise of the "rich" in America. "Yes, the rich keep getting richer," concluded one article while another article, bemoaning a growing "income gap," worried that "the promise of a higher standard of living is limited to a few at the top" and asked: "How much inequality can a democracy stand?"

Such concerns are misguided, for the much-publicized widening of a wealth/income gap in the United States is largely illusory. The gap is supposed to have widened particularly during the years of the Reagan presidency, when income tax rates were cut sharply. Yet, in those years (1983–89), the net worth in terms of the 1989 dollar of families earning $50,000 or more annually rose only 6.6 percent. The comparable increase for families earning from $20,000 to $29,999 was

28.9 percent and the increase for families earning from $30,000 to $49,999 was 27.7 percent.[1] Moreover, the net worth of whites rose more slowly (24 percent) in those years than for blacks (35 percent) or Hispanics (54 percent).[2]

There has also been much concern that the all-important "middle class" is somehow disappearing in America. Encouraging this misperception are the aforementioned data showing a decline in the real median income of U.S. households, as well as inflammatory rhetoric from ax-grinding politicians and academicians, such as Robert Reich, a professor at Brandeis University and President Clinton's former secretary of labor. Reich is fond of proclaiming that for a decade and a half, ordinary families have been working harder and getting less.

Nothing could be further from the truth. What has been happening is that, with the remarkable increase in wealth in America, we are witnessing an unprecedented event: the creation of history's first mass upper class. As David Frum, a senior fellow at the nonprofit Manhattan Institute observed, "Just as the American economy pulled its most talented people out of poverty in the 1950s, it pulled the ablest of its middle class into affluence in the 1980s."[3]

Evidence of this remarkable development may be found in a few numbers. Between 1980 and 1993, the percentage of U.S. households earning between $25,000 and $75,000, again in 1993 dollars to adjust for inflation, fell to about 47 percent from nearly 51 percent. But this reduction is not due to any mass slippage of middle-class families into lower income groups, for in the same period the percentage of households

---

1. Study by the Federal Reserve and Joint Economic Committee, reported in the *Wall Street Journal*, July 11, 1995.

2. Calculations by Federal Reserve Bank of St. Louis, reported in the *Wall Street Journal*, June 14, 1995.

3. David Frum, "Welcome, Nouveaux Riches," *New York Times*, August 14, 1995.

earning less than $25,000 changed little. Rather, the reduction reflects more and more families ascending to higher income categories.

Data documenting this rising affluence are striking. In the late 1960s, some 1 million U.S. households earned more than $100,000, in 1993 dollars. By the start of the 1980s, some 2.7 million households had crossed above the $100,000 mark. By the early 1990s, as many as 5.6 million households were earning over $100,000. This progressive increase in real income, of course, far outpaces the rise in population.

At present, about 1 million families have incomes exceeding $200,000, a statistic prompting David Frum to remark: "Nothing like this immense crowd of wealthy people has been seen in the history of the planet." The pattern, he contends, is merely "the latest manifestation of the great miracle of modern American life, the long 20th century tidal roll of ever-increasing abundance."

To the extent that a wealth/income gap does exist in the United States or elsewhere, I would simply note that a democracy in fact *needs* a considerable degree of inequality. As Michael Novak, the 1994 Templeton Laureate, observed in the *Wall Street Journal* (July 11, 1995): "Equality under law is one thing, and a good one, but in an inventive and dynamic society equal (even relatively equal) incomes can be achieved only by abandoning liberty for tyranny." As Novak says, "No nation that rewards effort, talent, inventiveness and luck can even pretend to cherish equal outcomes."

A Global Trend

The "tidal roll" of abundance that David Frum depicts is by no means limited to the United States. In Britain, for example, some 70 percent of the nation's families now own their homes, up from about 50 percent as recently as 1979. These homes, moreover, are becoming ever more comfortable. In 1980, less than half had a deep freezer, while now nearly 90 percent are equipped with these appliances. There has been a similar proliferation of air conditioning, particularly in the southern part of the country. And telephones, which were uncommon in British homes as recently as the 1960s, have become ubiquitous. In the 1970s, it took up to three months to have a phone installed, while today the job is usually done a couple of days after the request is made.

The wave of abundance is rapidly spreading beyond highly developed nations. It is sweeping around the world, even through much of the so-called Third World, nations once deemed to be inescapably mired in poverty. In India, a middle class more than twice the size of Britain's entire population is emerging, and it is expanding at a rate of about 20 percent a year. In China, with its growing private sector, an entrepreneur establishes a new business every eleven minutes in Shanghai alone. Thailand's real per capita GDP is rising at a remarkably swift pace—5.5 percent a year, on the average, since 1975.

By most measures, in fact, economic activity is rising appreciably faster within the emerging markets of the Third World than in developed nations such as the United States, Japan, and the West European bloc. Since the mid-1970s, World Bank data show, the inflation-adjusted rate of economic growth for emerging nations has averaged about 6 percent annually, or roughly twice the comparable rate for developed countries. China's growth in that period has averaged 8.2 percent annually and Indonesia's 6.4 percent. These rates compare with yearly growth of 3.6 percent in Japan, 2.3 percent in the United States, and 1.7 percent in Britain. To be sure, population is growing far more rapidly in emerging areas than in the developed world. Even taking this into account, however, economic advancement in the emerging countries is relatively swift. Since the mid-1960s, annual growth in real per capita GDP has averaged 3.4 percent in emerging nations, against only 2.5 percent in developed ones.

Though living standards in most developed nations remain substantially higher than in emerging ones, the relatively swift economic growth in these nations is bringing solid gains in living standards. These gains, described at greater length elsewhere in this book, take many forms. The average life expectancy at birth in emerging countries now stands at seventy years, according to the United Nations. That remains about eight years less than in developed nations, but as recently as 1955 the gap was sixteen years. Then, the average life expectancy in emerging countries was only fifty-three years, against sixty-nine years in developed ones.

There also are impressive advances in literacy rates in emerging nations. More than 80 percent of adults in the Third World are now literate, according to the World Bank, up from only 58 percent in 1960. In the same period, by comparison, adult literacy in developed nations has risen one percentage point, to 99 percent. The sharpest gains since 1960 have been in Indonesia, where adult literacy has jumped to 77 percent from 39 percent, and in Turkey, with a rise to 81 percent from 38 percent.

Also on the increase in emerging nations is the daily intake of calories. On a per capita basis, this intake has climbed 17 percent in the past three decades, compared with a 7 percent rise in developed countries. Among the sharpest increases: Indonesia, up 47 percent; and China, up 40 percent.

The number of physicians per thousand people—another barometer of living standards—has climbed at an impressive pace in emerging nations. Since the mid-1960s, this ratio has risen 136 percent, from 0.47 per 1,000 to 1.11. This compares with a gain of 57 percent in developed countries, from 1.22 per 1,000 to 1.92. Again, Indonesia's showing stands out, with an increase of 333 percent, followed by Brazil's gain of 265 percent.

Still other evidence of higher living standards in the Third World is the increasing availability of such conveniences as telephones and television sets. Since 1975, the number of telephone lines per thousand people in emerging countries has increased over 180 percent, more than twice the comparable gain in developed nations.

Just since 1980, the number of emerging-market households with television sets has risen more than 150 percent, while the comparable gain in developed countries is about 20 percent. In 1980, there were more than two U.S. households with TV for every Chinese household so equipped. Now, there are some five million *more* households with TV in China than in the United States. Similarly, Brazil now has more TV households than Japan, and Indonesia has as many as Britain. In India, still poor despite a growing middle class, sales of color TV sets surged 18 percent to 1.4 million units in a recent twelve months and sales of refrigerators rose 15 percent. Between 200 million and 300 million people, roughly one-third of India's population, regularly buy mass-produced consumer goods nowadays.[4]

Looking Ahead

For all the gains in living standards in emerging nations, a vast potential of unfilled needs remains. In Mexico, only 31 percent of households have clothes dryers and only 19 percent have microwave ovens. In the United States, by comparison, 74 percent of households have clothes dryers and 44 percent have microwaves. And Mexico, I should add, is among the more highly developed of the emerging nations.

Demographics also point to burgeoning demand in emerging markets. Not only do some 85 percent of the world's

4. Sally D. Goll, "India's Growing Middle Class Purchases Stuff Dreams Are Made Of," *Wall Street Journal*, July 7, 1995.

population of 5.5 billion reside in these nations, but the age structure of this group is appreciably younger than that of developed nations. In India, some 64 percent of the population is under the age of thirty. In Indonesia, the percentage is 65 percent. The comparable rate in the United States is 46 percent and in Japan only 40 percent.

Looking ahead, concern is often voiced in the United States that people born shortly after World War II—the so-called baby boomers—face a bleak financial future. The baby boomers appear to be saving perilously little, the doomsayers warn. Moreover, they note, the percentage of the U.S. population of working age inevitably will shrink as more and more baby boomers retire, eventually bankrupting the Social Security retirement system.

Overlooked in all this pessimism, however, is the highly encouraging fact that the baby-boom generation stands to benefit hugely from transfers of wealth from older Americans. In 1992 alone, parents and other relatives gave some $4.4 billion to help baby boomers buy homes, according to one survey. Another finds that in 1986 parents transferred nearly $100 billion to help with educational and other expenses. One household in every four surveyed shared at least some wealth with its adult children and one in eight shared wealth with its grandchildren.[5]

As more and more parents of baby boomers die and leave their assets to heirs, this unprecedented transfer of wealth from one generation to the next will continue to swell. Recent moves in Washington to reduce the inheritance tax burden of many U.S. families will further strengthen the financial position—and ultimately the living standards—of the baby-boom generation and the generations to follow. As the twenty-first century unfolds, our children and our children's children will live lives of unprecedented comfort and wealth.

5. G. Pascal Zachary, "Research Says Transfers Bolster Living Standards," *Wall Street Journal*, February 9, 1995.

2.

Food

"First feed the face, then talk right and wrong."
—Bertolt Brecht (1898–1956), *Three Penny Opera*

- As recently as 1940, one American farm worker produced enough food to supply almost a dozen people, while now a single worker turns out enough food for about eighty people.
- The world's population now consumes about 20 percent more food per person than fifty years ago.
- People in developing countries now consume, on average, 2,600 calories per day, more than 10 percent *above* the required minimum.
- Since 1950, the value of U.S. farm output has climbed nearly six-fold, after allowing for inflation.

From Scarcity to Plenty

While man does not live by bread alone, neither can he survive without food, which like air and water is essential to life. For primitive people, simply obtaining enough food to sustain themselves consumed almost all their time. They survived by scavenging for roots, nuts, berries, and grains and snaring small animals.

As humankind has evolved through the ages, we have learned how to grow our food, and there surely is time left over for talk of right and wrong. Nonetheless, the matter of sustenance has remained a major concern. In 1776, when the great Scottish economist Adam Smith published his classic *Inquiry into the Nature and Causes of the Wealth of Nations*, some 85 percent of the world's population was engaged—most inefficiently—in producing food. Even so, famines were widespread. According to the *Statistical Abstract of the United States (SAUS)*, as recently as 1929, some 10.4 million U.S. workers, more than one-fifth of the nation's entire work force, labored on farms. Today, in startling contrast, less than 3 percent of the U.S. work force is so engaged, and yet there is sufficient food produced to supply all Americans, with plenty left over to help feed others around the world.

To be sure, television cameras are quick to focus on the skeletal frames and distended stomachs of famished people in such poverty-ridden areas of the less-developed world as Somalia, Sudan, and Ethiopia. Even in highly industrialized countries, as the media hasten to point out, people still go to bed hungry in circumstances that often seem more suited, say, to eighteenth century France than twentieth century America.

But such misery seems largely to reflect misallocations of supply and political roadblocks and repression, rather than any real insufficiency of food in the world. Indeed, the overriding truth is that today, by and large, the world's population of nearly 6 billion people is better fed than ever before. Never has food been so plentiful or so nutritious, or been available in such remarkable variety—a situation made possible by the advent of highly sophisticated agricultural equipment, large-scale commercial farming, new cultivation techniques, improved fertilizers and pesticides, and genetically engineered plants and even animals. "The world has more food security than at any time in history," says Dennis Avery, head of the Center for Global Food Issues at the Hudson Institute in Indianapolis, Ind.[1]

Nowhere has the progress been more pronounced than in the United States. Only fifty years ago, again according to the

---

1. Barnaby J. Feder, "Shortfall In the Grain Fields," *New York Times,* November 19, 1995.

Marie Antoinette on her way to the guillotine, October 1793. Parisians, starving and without bread to eat, were not amused when Marie Antoinette suggested they eat cake instead.

SAUS, there were some 6 million farms in America and now there are fewer than 2 million. However, today's farms cover about as much acreage as was in operation at the end of World War II. The average size of a U.S. farm now is about 500 acres, up from about 190 acres a half century ago. Larger in size than its predecessors, today's farm is vastly more productive. Just since 1950, the value of U.S. farm output has climbed nearly six-fold, after allowing for inflation. As recently as 1940, one U.S. farm worker produced enough food to supply almost a dozen people, while now a single worker turns out enough food for about eighty people.

Two startling statistics tell a similar story: Agricultural methods in place only thirty years ago—if used now—would require additional arable land the size of North America to provide the world with the same amount of food currently produced with modern methods. And feeding today's world population with techniques used as recently as the middle of the last century would require some 120 million square miles of land, or more than double the amount of land, including mountains, swamps, deserts, and frozen tundra, that exists on earth!

U.S. Farm Population

Year	Millions	Percent of Population	Number of Farms
1910	32.1	35	6,366
1920	31.9	30	6,454
1930	30.5	25	6,295
1940	30.5	23	6,102
1950	23.0	15	5,388
1960	15.6	9	3,962
1970	9.7	5	2,954
1980	7.2	3	2,440
1990	4.6	2	2,146

SOURCE: *Statistical Abstract of the United States*

The world's population now consumes about 20 percent more food per person than a half century ago. Just twenty-five years ago, people in developing countries were able to obtain, on average, only 80 percent of their minimum daily dietary needs, but now they receive, on average, nearly 2,600 calories a day, or more than 10 percent *above* the required minimum.

Thomas R. Malthus, an English clergyman and economist born in the latter half of the eighteenth century was an early prophet warning that the world's food supply would prove insufficient as its population kept expanding. He forecast that population would increase geometrically while food supplies would rise only arithmetically. The upshot, he was convinced, would be widespread poverty and starvation. This would happen, he warned in his alarming *Essay on Population*, even if "by great exertion, the whole produce of the [British Isles] might be increased every 25 years by a quantity of subsistence equal to what it at present produces." Dolefully, he added, "The most enthusiastic speculator cannot suppose a greater increase than this."

In fact, the "quantity of subsistence" around the world has been approximately doubling every fifteen years in recent decades, a situation that the pessimistic Reverend Malthus could not possibly have imagined. Nor would he have dreamt that two centuries after his *Essay* first appeared Britons would be dining routinely on such then-unfamiliar items as pineapples from Thailand, okra from Kenya, and avocados from Florida.

As recently as 1970, English food shops sold only locally grown vegetables that happened to be in season. Britons' vegetable diet in the autumn consisted mainly of cabbage, varieties of root vegetables, and Brussels sprouts. Less than fifty years ago, the choice of food in many British homes was limited in winter to a few vegetables, such as onions and potatoes, and a few fruits, such as apples, stored in a cold closet, called a larder. Meat and fish, of course, were preserved by salting and drying.

*Portrait of Thomas R. Malthus (1766–1834). Engraving by J. Linnell, ar.
1830.*

In British homes today, by comparison, there is an array of food from the European continent, much of it *trucked* to the island nation, as well as from such more-distant places as Africa, Asia, and the Americas. Mass production, swift transportation, canning, and freezing bring a year-round supply of fruits and vegetables. In Britain, as in many nations, frozen and canned foods take up entire aisles in supermarkets. And thanks to jet-propelled air transport from different climates around the world, fresh vegetables and fruit are available at any time.

For example, during Christmas week in Maidenhead, a commuter town thirty miles west of London, a shopper in a Waitrose supermarket nowadays may choose from an international array of fresh fruits. There are pineapples from Costa Rica, strawberries from Egypt, lychees from South Africa, cherries from Chile, passion fruit from Kenya, cherimoya from Spain, carambolas from Malaysia, Galia melon from Israel, and fresh figs, mangoes, and melons from Brazil. The vegetable bins are laden with sugar snap peas from Kenya, artichokes from Egypt, avocados from Israel, rettich from Italy, and varieties of asparagus from Chile and Thailand. For stir-fry addicts there are snow peas from South Africa, and ginger and shitake mushrooms from Thailand.

Nor is this English supermarket unique. Supermarket chains in Switzerland, France, and the United States offer equally exotic fare. Food, in fact, has become an exciting adventure for the millions of people in dozens of countries who now regard it as far more than simply "the staff of life." *The Packer*, a U.S. trade publication for the produce industry, finds that nearly one out of every four shoppers in America now experiments each year by purchasing some type of fruit that they have never tasted before.

Americans, like Britons, have available an unprecedented variety of ethnic cuisines. In 1955, a young American couple who previously had lived in Japan was dining in the front of a Chinese restaurant in McAllen, Texas. As they ate, they were startled to note that a crowd had gathered at the window to stare in at them. The couple eventually realized this curiosity was caused by their eating with chopsticks! Today, of course, Americans of every background often use chopsticks when they choose to dine at Korean, Japanese, Thai, or Chinese restaurants located in the United States.

The variety of food available nowadays is awesome. A dozen years ago, for example, man was able to cultivate only one kind of mushroom. Today, supermarket bins bulge with such commercially grown mushroom varieties as crimini, oyster, shitake, and portobello. As varieties of food multiply, they increasingly are being refined to be more health-giving. Cholesterol and fat are being removed from such items as mayonnaise, margarine, and sour cream. There are "lite" beers and "nonalcoholic" beers and ice cream that comes free of all fat. Low-fat and fat-free products now make up more that half the items on supermarket shelves in the United States. All packaged food sold in the United States now must list the amount of fat, cholesterol, and sodium contained in one serving. At the same time, health-conscious Americans continue to cut their intake of fat and cholesterol by trimming their consumption of such items as eggs and switching from regular to low-fat varieties of milk.

Spurring Output

In 1857, in a statement of social protest, Jean François Millet painted *The Gleaners*, portraying women stooped over picking up single grains of wheat left behind by the reapers.

Reaping wheat itself remained a manual task in many places long after Cyrus McCormick, an American, invented the horse-drawn mechanized reaper in 1831. Today, in contrast, such farming tasks are largely automated. Machines pick and shuck corn twelve rows at a time and automatically

The Gleaners *by Jean-Francois Millet. The Louvre, Paris.*

harvest such edibles as lettuce, peas, and tomatoes. A further spur to production is the development of super hybrids and genetically engineered items, such as monster-size pumpkins.

Not long after India's independence in 1947, doomsayers predicted that the world's second most populous nation would suffer devastating famines and be compelled to depend permanently on imported food to sustain its teeming millions. But in the early 1970s, Norman Bourlag, working at the International Maize and Wheat Improvement Centre in Mexico, developed a revolutionary strain of high-yielding dwarf wheat. This single development ushered in the so-called green revolution that transformed India, among other nations, from a recipient of massive food aid into—*mirabile dictu!*—a net exporter of food, according to the United Nations Development Programme (UNDP). In 1970, more than 20 percent of

Nathan, left, and Paula Zehr pose with their 1,061-pound monster pump-kin that won them $50,000 at the World Pumpkin Confederation competition Saturday, Oct. 5, 1996 in Clarence, N.Y.

India's merchandise imports were food products, compared with only 5 percent today, World Bank statistics show.

Meanwhile, research efforts supported by the Consultative Group on International Agricultural Research, a non-profit group in Washington, have led to the production of high-yielding staples that resist disease and damage by various pests. Similarly, researchers at the International Rice Research Institute in the Philippines have developed varieties of "dwarf" rice whose short, thick stalks support heavier grain loads and therefore will not topple into water-filled paddies. This has brought a doubling, and in some cases a tripling, of rice yields in many areas.

Indeed, such breakthroughs have allowed the per capita

output of rice and wheat, key staples for two-thirds of the world's population, to rise over the past three decades, despite a population increase of nearly 2 billion. Still more encouraging is that these staples sell for only about half of what they did in the late 1960s.

Nutritional Gains

The nutritional quality of most staples is steadily improving as well. Using germ plasm, or the reproductive cells, from high-quality corn supplied by the International Maize and Wheat Improvement Centre, planters in Brazil recently developed a so-called BR-451 strain. The strain-contains two essential amino acids that provide greatly increased nutritional value for animals as well as humans.

In addition to genetically-based nutritional improvements, researchers are finding new ways to protect plants from the diseases and insects that plague them and frequently destroy large quantities of crops. A new advance in genetic research promises, for example, to protect rice plants

Rice and Wheat: More to Eat at Lower Prices

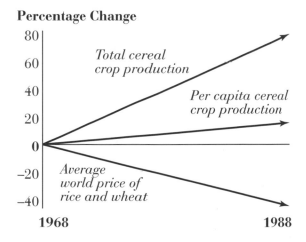

SOURCE: United Nations Development Programme

from leaf blight, a disease that in some years has destroyed up to half the rice crop in some parts of Asia and Africa. The protective gene, called Xa21, is a major step toward increasing the world's food supply: it marks the first disease-resistant gene ever isolated in rice. Although it does not exist in all varieties, there is hope that scientists will manage eventually to protect all rices, as well as other grains and vegetables, in a similar fashion.

Additionally, the International Maize and Wheat Improvement Centre is making tropical maize more resistant to its primary insect pest, the stem borer. In cooperation with scientists in Mexico and Brazil, it is injecting into the plant the genes of *Bacillus Thuringiensis*, a protein that is toxic to the pest.

The ability to protect plants genetically by manipulating their immune systems will be crucial in efforts to raise further the world's food supply, since there are limits to and environmental concern over the use of fertilizers and pesticides. Indonesia, for example, had sharply boosted its rice production through the heavy use of fertilizers and pesticides when suddenly the output began to fall off due to infestations of the Brown Plant Hopper, a pest that feeds on rice leaves. Additional pesticide applications only worsened matters by destroying the natural enemies of the Brown Plant Hopper, while allowing a mutated strain of the pest to proliferate. According to the UNDP, the Indonesian government subsequently banned some sixty pesticides used on rice and, as an alternative, began encouraging farmers to plant high-yielding, pest-tolerant rice supplied by the International Rice Research Institute. As a result, rice production rose sharply and, as a bonus, over $150 million a year was saved—money that otherwise would have gone for pesticide purchases.

Scientists are also homing in on ways to destroy pests themselves. In mid-1988, the screwworm fly, a pest that had devastated livestock in the Western Hemisphere, suddenly began ravaging sheep and goats in Libya. The menace threatened the livelihood of millions of Arab and African farm families dependent on livestock. An international effort was mounted that involved releasing hundreds of millions of flies rendered sterile by radiation. When the sterile males mated with wild females, the population died out.

Researchers are also working on changing the genetic structure of the Mediterranean fruit fly in an effort to halt the hundreds of millions of dollars of damage that the pest inflicts on crops annually. In addition, genetic research is under way to combat the desert locust, which has destroyed crops for more than four thousand years, eating its own weight in vegetation every twenty-four hours. One approach is to alter the gene that causes the locust to swarm.

Researchers are also working to improve the quality and quantity of the animals we eat. Fish constitute the main source of animal protein for more than 1 billion people around the world, and further increases in population and income are expected to boost worldwide demand for fish by at least one-third over the next few years. Yet, the natural supply of fish in oceans, lakes, and rivers has been exploited and is severely depleted. Even so, more than offsetting this unfortunate situation is the increasing cultivation of fish by commercial means—fish farms, if you will. This cultivated protein source has been expanding at the remarkably brisk rate of about 7 percent annually.

A new breed of a type of fish called tilapia, which matures twice as quickly and is twice as big as its natural African forebear of the same name, has been developed in the Philippines. This new breed of farmed fish has been labeled the "aquatic chicken" and serves as a valuable protein source for low-income families throughout Southeast Asia.

In the United States, meanwhile, poultry growers using new breeding techniques have managed to increase sharply the amount of breast meat on chickens and turkeys. Breeder

turkeys, which are now nearly twice the size they were thirty-
five years ago, have such overblown breasts that they have
difficulty standing and are so ungainly that most must be
bred through artificial insemination. One commercial breeder
recently managed to produce an 86-pound turkey, more than
four times the average size. The offspring of approximately
ten thousand "super" turkeys are responsible for producing
more than 90 percent of all Christmas and Thanksgiving
turkeys in the United States, as well as ground turkey, turkey
hot dogs, and other processed meat products. The breasts of
their smaller progeny also vastly overshadow the drumsticks
they stand on, catering to the fact that two of every three
Americans prefer white meat to dark.[2]

Another genetic breakthrough is the Flavr Savr, a high-
tech tomato. Many tomatoes found in American supermarkets are picked when they are hard and green. When they are
artificially inspired to ripen with ethylene gas, they lose
much of the flavor of a vine ripened tomato. But the Flavr
Savr tomato is injected with a gene that allows it to ripen
more slowly on the vine, thus retaining flavor. The slower
ripening process, moreover, cuts loss through spoilage. Fully
half the vegetable crops produced in the United States are
lost to spoilage during shipping, on the store shelf, or in the
refrigerator.

Thanks to such innovations as the Flavr Savr, food
around the world is cheaper, tastier, and more nutritional
than ever before. In many cases, prices are lower than years
ago, even when measured in terms of today's vastly cheaper
currencies. In the United States, chicken prices, for instance,
have declined substantially, even without taking inflation
into account. A quarter century ago, the Shoprite supermarket chain was advertising a "special" price of 77¢ a pound for

2. Mark Fritz, "Breeders Rule the Roost," *The Record,* November 19, 1995.

chicken legs, or 28¢ *more* per pound than the price of a similar, recent Shoprite special.

Food in the Future

While it is difficult to understate the role in recent decades of technological innovation, an enormous spur to still larger food supplies in coming years will, I believe, result from advances on the political front. Increasingly, nations that once espoused state-ownership are privatizing the agricultural sectors of their economies. The motivating idea is one that Adam Smith would have readily endorsed—that farmers will work harder and more efficiently if they are truly toiling for themselves, rather than for faceless governmental bureaucracies.

Already, there is evidence that shifts from public to private ownership are serving to spur farm output. Consider the case of Poland, a formerly Communist nation. Polish farmers once were compelled to sell their produce to state enterprises, as well as to buy such farm inputs as seeds, fertilizer, and tractors from the state. Their costs and prices were generally determined by the state, rather than the marketplace. Now, in contrast, Polish agriculture is largely privatized, and how the nation's farmers run their operations is largely up to them and the markets, and no longer under rigid state control. The upshot is a substantial increase in Poland's agricultural output.

As more and more of the world's farming is privatized, as I am confident will happen, global agricultural output should expand accordingly. After all, whether we are talking about bricklayers or farmers, people generally work harder and better when they know they are working for themselves and can reap rewards for their labors. An indication of this tendency was the remarkable success of so-called land reform in various developing nations soon after World War II. This involved the redistribution of ownership through subdividing large

3.

Health and Life Expectancy

"Healthy citizens are the greatest asset any country can have."
—Winston Churchill, radio broadcast, March 21, 1942

- Fifty percent of all medical knowledge has been developed in the past decade.
- The U.S. spends $15.9 billion annually on medical research, 5,300 times the 1940 level.
- American doctors perform nearly 2,000 heart transplants, 9,000 kidney transplants, and 40,000 cornea grafts per year.
- The U.S. death rate from heart diseases has dropped 40 percent in the last two decades.
- New drugs are coming to the American market at the rate of approximately one every two weeks.

Declining Diseases

In 1981, NBC-TV weatherman Willard Scott came up with the idea of wishing people turning one hundred years old a happy birthday on the air. But what started out as an occasional event, with few qualifiers, has turned into an everyday feature, with some four hundred new centenarians a week vying for Willard's airtime.

Around the world, as this remarkable competition suggests, people are healthier and living longer than ever before. In the United States, average life expectancy at birth is now about seventy-five years, double what it was two centuries ago, according to the *SAUS*. In the Third World, the average now exceeds sixty-three years, up from less than thirty at the start of this century. In China, it approximates seventy years, up from only twenty-four as recently as 1930.[1]

To put this remarkable increase in worldwide life expectancies another way, nearly two-thirds of all the people ever born—an estimated 80 billion—who lived beyond the age of 65 are alive today.

In the industrialized world, centenarians are the fastest growing age group. Their number has approximately doubled every decade since 1950. In the United States, some 52,000 people are alive who have passed their one hundredth birthdays, triple the number as recently as 1980. More remarkably, fully half of these people are in good physical and mental health, with many holding down active jobs. [2]

The world's population, at less than 500 million, didn't change much for about three thousand years, up to the start of the fifteenth century. But then a rise began which, by the middle of the nineteenth century, had become almost vertical, as seen in the chart at the top of page 51. Since then, the world's population has surged from less than 1 billion people to some 6 billion. Sweeping advances in medicine, surgery, medical equipment, living conditions, and nutrition account for this dramatic increase.

Scientific and medical discoveries have sharply reduced and even occasionally eliminated many parasitic and infectious diseases that have long plagued humankind. This list includes bubonic plague, smallpox, measles, typhus, scarlet fever, polio, and cholera. A steep drop in the incidence of these diseases in just seventy-five years is evident in the table opposite. Keep in mind that the U.S. population, now above 260 million, has more than doubled since 1930.

One amazing story is that of the eradication of smallpox, a disease that wiped out entire communities in the 1700s. In

1. Julian L. Simon, *The State of Humanity.*
2. Caryl Stern, "Who is Old?" *Parade Magazine*, January 21, 1996.

Estimation of World Population, 1600 B.C. to 2000 A.D.

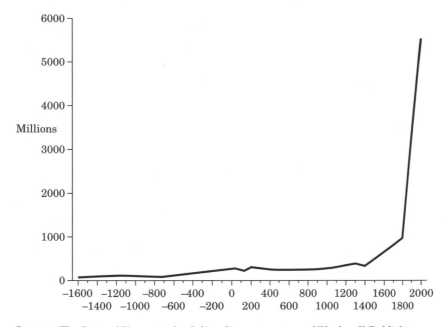

SOURCE: *The State of Humanity* by Julian Simon, courtesy of Blackwell Publishers, Oxford, England

Cases of Infectious and Parasitic Diseases Reported in the U.S.

	1930	1970	1994
Smallpox	48,700	0	0
Measles	418,500	47,400	301*
Mumps	N/A	105,000	1,500
Malaria	98,200	3,051	1,229
Whooping cough	166,500	4,200	4,600
Diphtheria	66,400	435	2
Polio	9,200	33	0

* This 1995 figure from the Centers for Disease Control and Prevention represents an all time low for the disease.

SOURCE: *Statistical Abstract of the United States*

the late eighteenth century, Edward Jenner, an English doctor, noticed that dairy maids seldom contracted smallpox. In 1796, he inoculated a young boy with cowpox, a common disease among milkmaids, which successfully immunized him against the much more deadly smallpox. By 1977, following a ten-year campaign by the World Health Organization, smallpox was wiped out as a disease in every country in the world. The United Nations recently recommended destroying by mid-1999 the last laboratory stocks of the virus, which are held by the United States and Russia.

Polio, which has been around as long as the pyramids, is on the verge of being wiped out worldwide. The worst epidemic in the United States occurred in 1952, when a record 58,000 cases were reported. Two years later, Jonas Salk produced a polio vaccine and three years later a live vaccine was developed.[3] By 1969, not a single death from polio was reported in the United States. As a result of polio's decline, the March of Dimes, founded in 1938 to combat polio by President Franklin D. Roosevelt, himself a polio victim, was able to redirect its main mission to fighting birth defects.

Inoculations against typhoid, tuberculosis, whooping cough, tetanus, and flu have all been developed in the last one hundred years. And those against mumps, polio, and measles have all been developed in the last fifty years. I should add, in this regard, that experts estimate that some 95 percent of all medical knowledge has been developed in this century and fully 50 percent in just the past decade.

Huge Gains in Child Survival

Chances of children surviving their first year have improved dramatically. Less than two centuries ago, only 800 children

---

3. Harold M. Schmeck, Jr. "Dr. Jonas Salk, Whose Vaccine Turned Tide on Polio, Dies at 80," *New York Times*, June 24, 1995.

Dr. Edward Jenner giving a child a vaccination depicted in this sculpture by G. Monteverde.

in every 1,000 born in Sweden, for example, survived more than a year. By 1950, the survival rate was 980 out of 1,000 and today it stands at 995. In the United Kingdom since 1979, the proportion of babies dying in their first year has dropped by almost half.

Especially remarkable strides have been made in developing countries. Early this century, only about 70 percent of all babies born in Chile survived their first year, which was a main reason why Chilean life expectancy at birth averaged 30 years. Today, in contrast, 98.4 percent of Chilean children live past their first year and life expectancy is about 74 years. The pattern is evident even in much poorer developing nations than Chile. In Gambia, one of Africa's poorest countries, 87 percent of children now survive beyond year one, up from only 57 percent as recently as 1950.

Huge steps are being taken in the developing world to reduce childhood diseases. WHO's Expanded Program on Immunization has targeted six major childhood diseases: diphtheria, tetanus, tuberculosis, polio, measles, and whooping cough. Through the program, the percentage of the world's children immunized against these diseases has jumped from 5 percent to 80 percent in less than two decades.[4]

Additionally, the UN-sponsored Programme for Vaccine Development (PVD) supports projects around the world to develop new vaccines against still unchecked major viral and bacterial diseases. It also backs efforts to produce vaccines that don't require refrigeration, are more stable in tropical temperatures, and need only one shot, with no boosters.

Many diseases that cause severe illness in young children living in low-income countries are brought on by parasitic worms. Research shows that nineteen of twenty-three major worm-caused diseases can be controlled by three drugs: Albendazole, which kills round, whip, and hook worms;

4. "U.N. Reports Success in Children's Vaccines," *The Record,* October 1, 1996.

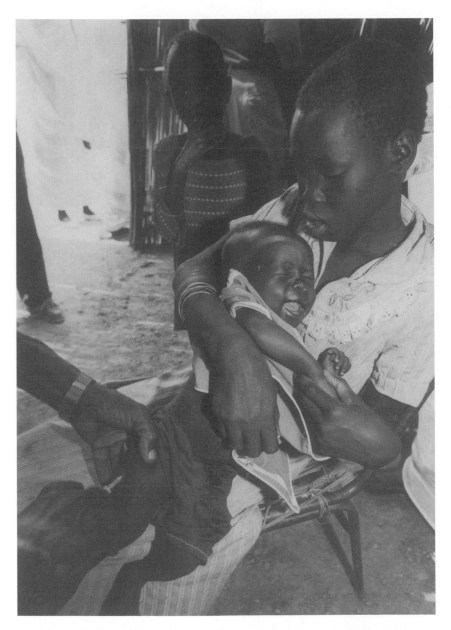

A Sudanese child cries in his mother's arms while being vaccinated against measles in a medical unit in Bantiou near Khartoum.

Ivermectin, which fights filarial worms; and Paraziquantel, which controls schistosome worms. The United Nations Development Programme, UNICEF, WHO, and the Rockefeller Foundation are supporting large operational field trials to distribute these drugs, along with iodine, iron, and vitamin A, through primary schools in poorer countries. Using teachers to give the medication and nutrients allows for far greater coverage than could be achieved by health workers alone.

At the same time, UN and U.S. agencies, working with volunteer groups and some major corporations, have launched a major campaign against two debilitating worm diseases prevalent in Africa. One is the Guinea worm, which comes from tiny larvae found in untreated drinking water. The worm mates inside the human body and emerges a yard-long through the skin. Less than a decade ago, some 20 million people in the developing world were afflicted with this painful disease. Today, it is on the verge of eradication, since the larvae can be killed by straining or boiling the water, or by applying a larvicide developed and supplied free by American Cyanamid Co. In a related act of corporate good will, DuPont & Co. recently teamed with Precision Fabrics Group, to donate 1.4 million reusable nylon filters to strain the parasite from drinking water.

Another parasitic worm under attack causes onchocerciasis, or river blindness, to people bitten by the black fly. Some 17 million people in eleven West African countries have been infected and possibly blinded. But in the last decade a medication has been developed that kills the larvae produced by the worm. Merck & Co. provides the drug, called Ivermectin, free throughout West Africa. These donations already total about $250 million.

Malaria, which struck over 180,000 Americans a year in the early 1920s, today affects less than 1,300 a year, though the population has grown about 150 percent in the interim. To be sure, in some developing countries, such as Thailand, malaria remains a serious threat, due largely to drug-resistant strains of the mosquito-borne disease. Even in such places, however, progress is being made through such basic measures as the increasing use of netting impregnated with Permethrin, an insecticide. In a recent ten-month period, for instance, 10,000 such nets were supplied to a district of the Central African Republic occupied by about 40,000 people.

Modern Medicine

According to the *SAUS*, there are more doctors and nurses serving the American public now than ever before. Since 1970—a period in which the U.S. population has increased by about one quarter—the number of doctors working in the United States has roughly doubled, to more than 600,000, and there are more than three times as many registered nurses.

In 1940, the United States spent less than $4 billion on health, and only $3 million on medical research. Today, the nation invests $950 billion annually, or nearly 14 percent of its gross domestic product, on health and its spending on medical research now comes to $15.9 billion, or some 5,300 times the 1940 level.

By no coincidence, the number of hospitals, the number of beds per hospital, and the occupancy rates have all been gradually, but steadily, declining in the United States since 1972—under the circumstances, a most heartening trend.

In England, the number of general practitioners grew by more than 11 percent between 1984 and 1994, so that case loads were reduced by about 9 percent, allowing doctors to spend more time with individual patients. In addition, spending on the National Health Service in England has grown by 66 percent in real terms since 1979.[5]

5. Conservative Party Position Paper, January 19, 1996.

Even in the fight against cancer, there has been encouraging progress. In the United States, the five-year survival rate now exceeds 50 percent, up from only 20 percent in 1930. A major factor in this trend has been the establishment of a direct link between smoking and lung cancer, which has led in turn to a sharp decline in the number of Americans who smoke. In 1965, more than 42 percent of the adult U.S. population did so, while now only 25 percent do.

The recent discovery of so-called oncogenes, genes that trigger the spread of cancer, represents another big advance in the battle. As a result, researchers are working to develop drugs that will shut down oncogenes that become activated.[6]

A large benefit of a healthier worldwide population is that workplace absenteeism is dropping in many nations. In the Netherlands, illness-related absences have dropped from 17 percent of time worked as recently as 1983 to about 7 percent. In France in the same period, the absentee rate has fallen from 15 percent to 8 percent, and in Germany, from nearly 10 percent to under 7 percent.

Awesome gains have been made in the field of surgery, a procedure that dates back to the Stone Age, judging from the remains of prehistoric amputees. In fact, surgery was once held in such low regard by physicians that they handed over the task to barbers. As recently as the middle of the last century, surgical patients were often given a mixture of alcohol and gunpowder to drink. Then they would clench a piece of pipe between their teeth while friends held them down during operations. These were done as speedily as possible and often involved cutting things off and then cauterizing the stump with a red-hot iron.

Today's surgeons, in dazzling contrast, can miraculously reattach body parts, including severed fingers, hands, feet,

6. Earl Ubell, "What Medicine Will Conquer Next," *Parade Magazine*, November 5, 1995.

and even private parts. And, of course, organ transplants are adding precious years to the lives of hundreds of thousands of people.

It was in late 1967 that Christiaan Barnard, a South African surgeon, performed the first heart transplant. Within a year, the procedure was extending the lives of people with severely diseased hearts. Then, in 1982, the first artificial, electrically-powered heart was implanted by Robert Jarvik. It kept the recipient alive for 112 days.[7]

According to the *SAUS*, doctors in America now perform nearly 2,000 heart transplants a year, more than 30 times as many as in 1981. Over the same period, the annual number of liver transplants has risen more than 120 times and kidney transplants have increased by over 80 percent to more than 9,000 a year. In addition, more than 40,000 Americans receive cornea grafts annually, up from 15,500 in 1981.

Other remarkable operations, besides transplants, play a major role in extending lives nowadays. Some 170,000 Americans undergo coronary artery bypass surgery each year. This operation, once considered exceedingly dangerous, is today performed almost routinely, prolonging lives of people with severely blocked coronary arteries. Another new surgical development, called angioplasty, treats atherosclerotic arteries by threading a catheter through the vessels and then inflating a balloon-like tip to clear blockages.

Meanwhile, beta-blocking drugs, which slow the heart rate, are countering high blood pressure. Thanks, in part, to these drugs, the American death rate from heart disease has dropped by 40 percent in the last two decades. Claude Lenfant, director of the National Heart, Lung and Blood Institute, believes that in the first half of the twenty-first century, heart disease and heart failure may be overcome.[8]

7. "British Patient Given Artificial Heart," *New York Times,* October 30, 1996.
8. Earl Ubell, "What Medicine Will Conquer Next."

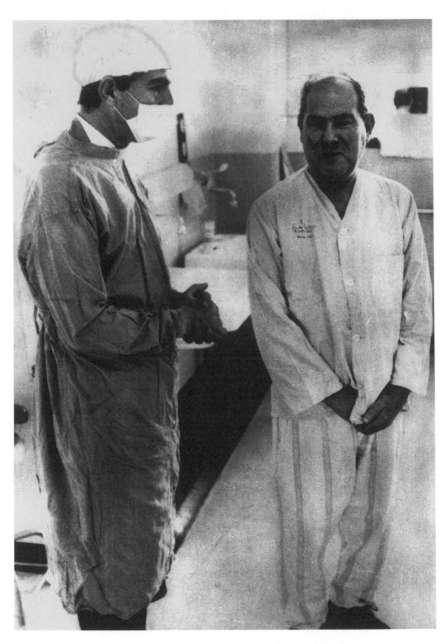

Posing for pictures for the first time since his heart transplant operation, dentist Philip Blaiberg (right) stands beside Dr. Christiaan Barnard at Cape Town's Groote Schuur Hospital, February 1968.

New medical technologies are contributing to life-prolonging diagnoses and cures. In early 1996, the U.S. Food and Drug Administration approved a machine that cleans LDL, a harmful form of cholesterol, from arteries of people severely affected by the vessel-clogging substance. This machine, called a Liposorber, removes blood slowly from a patient's arm and pumps it through a device that removes LDL, and then injects the cleansed blood back into the other arm.

At the same time, radioisotope scans have vastly improved the ability of doctors to detect malfunctioning or malignant organs. If an organ absorbs too much or too little of a radioisotope ingested by a patient, such a scan can tell the doctor that the organ is malfunctioning. Or, if a portion of the organ doesn't show up on the screen, this may indicate a tumor.

Computerized axial tomography (CAT) is a painless process employing complex, costly equipment to detect tumors, cysts, and abscesses that do not appear on less sophisticated diagnostic equipment. Developed in 1972, CAT scans can even sometimes distinguish between malignant and benign tumors. They are about one hundred times more sensitive than conventional x-rays, and don't pose the same radioactive danger.

Another painless, nonintrusive way of peering inside the human body is through magnetic resonance imagining, or MRI. This device surrounds the portion of the body being examined, emitting computer controlled radio waves and magnetic fields that are up to 25,000 times stronger than Earth's natural magnetic field. MRI enables doctors to receive three-dimensional images of body tissues.

Doctors are relying increasingly on ultrasonography, a relatively inexpensive imaging technique that uses ultra-high-frequency sound waves to locate and visualize internal organs composed of soft tissue that don't x-ray easily. Ultrasound technology is also being used to treat certain heart defects in fetuses.

Doctors are also increasingly exploring inside patients with cables made of fiberglass threads that transmit light and can be guided around bends in the organs and vessels. Called endoscopy, the procedure enables physicians to check for cancerous polyps, as well as perform delicate surgery on joints and biopsies on internal organs.

Another relatively new life-saver is a test called the prostate-specific antigen, or simply PSA. It enables laboratories to detect possible prostate cancer from small samples of blood. Treatment of prostate cancer, moreover, has been refined by brachytherapy, or the implantation of radiated seeds, which are only the size of a grain of rice, directly into the male sex gland. As effective as surgery in arresting prostate cancer, the technique marks a huge improvement on external beam radiation, which can damage normal tissue as well as attack cancerous tissue.

Elsewhere on the surgical front, gall bladders now can be removed through just a small incision in the abdomen, an operation that only a few years ago required opening the entire stomach, which in turn necessitated weeks of painful recuperation. The new method is made possible by a device called a laparoscope, which, when inserted through a small incision in the abdomen, allows surgeons to peer inside.

While modern medicine has made great strides in surgical and nonsurgical treatments of diseases, it has also developed a vast array of drugs to combat many previously untreatable ailments. In fact, treatment of diseases with chemicals was unknown until 1910, when the German scientist Paul Ehrlich developed Salvarsan, a chemical that kills the bacteria that causes syphilis. And it wasn't until 1928 that the Scottish scientist Alexander Fleming accidentally discovered the first antibiotic, penicillin. Today, new drugs, after tough testing by the Food and Drug Administration, are coming to the U.S. market at the rate of about one every two weeks.

In 1995 alone, the FDA approved twenty-eight new drugs, including drugs to fight cancer and AIDS. One such drug is Epivir, also known as 3TC, which can stall the growth of the AIDS virus for months and even years. Epivir is particularly effective when combined with AZT, an earlier AIDS drug. The combination boosts the immune system and lowers the amount of HIV virus in the blood of AIDS victims.

Also approved in 1995, Fosamax is another important new drug that strengthens bones. It helps fight osteoporosis, a bone-weakening disease that affects about one-third of all women after menopause. In clinical trials, the drug reduced the risk of hip and spinal fracture by about 50 percent among women with previous spinal fractures.[9]

Other recent breakthroughs in medication include drugs to treat such afflictions as diabetes, epilepsy, and Parkinson's disease. Experts are also predicting that in the years just ahead more than two dozen new drugs to treat strokes will be developed. Strokes afflict some 500,000 Americans each year and cause about 150,000 deaths, but the good news is that the death rate from strokes has fallen by 25 percent over the last decade.

Dentistry is another area where immense strides have recently occurred. In ancient China, "dentists" trained for the job, which consisted mainly of extracting teeth, by pulling nails out of boards with their fingers. Today, happily, extraction plays a much smaller role in dentistry, and of course is done less painfully. Now, most people's teeth last a lifetime, whereas only fifty years ago false teeth were commonplace among people over forty years of age.

Widespread use of fluoride in drinking water has played a major role in protecting teeth, while gum disease is thwarted by tartar-fighting toothpastes containing peroxide, baking soda, and fluoride. As recently as 1950, in contrast, fluoridated water was available to only about 1 percent of the U.S. population and there were no such toothpastes.

9. "Drug Shows Promise in Severe Osteoporosis," *The Record,* September 11, 1996.

At 154,000, the number of dentists in America has surged more than 60 percent just since 1970, according to the *SAUS*. Dentists are saving teeth with caps, bridges, and root canal surgery. It is even possible now for dentists to implant new artificial teeth in a patient's jaw. Bridges that were once made of porcelain arc now made of epoxy resins and cured to the proper color in the mouth. Similarly, cavities today are filled with an epoxy that lasts longer, is safer, and carries more cosmetic appeal than the metal amalgam that was used for decades. Meanwhile, high-speed, water-cooled drills and bloodless ultrasonic cleaning devices have made the dental chair less terrifying, and probably as a result more people are appearing for preventive checkups. Scientists, moreover, are experimenting with electrical currents, too weak to be felt by patients, to locate tooth decay long before it is visible with dental x-ray techniques.

New Trends in Medicine

In the last few years, scientists have been working on gene therapy, seeking to treat disease with the body's own genetic material. The theory behind gene therapy is that by inserting corrective genes into a patient's cells, genetically-based diseases such as hemochromatosis and cystic fibrosis may be eliminated.

"Twenty years from now, gene therapy will have revolutionized the practice of medicine," according to Dr. W. French Anderson, director of the gene therapy program at the University of Southern California/Norris Comprehensive Cancer Center in Los Angeles. "Virtually every disease will have gene therapy as one of its treatments."[10] To further this effort in

---

10. Leon Jaroff, "Keys to the Kingdom," *Time* Special Issue, "The Frontiers of Medicine," Fall 1996.

the United States, the National Institutes of Health are investing some $200 million a year in gene therapy research.[11]

To guide researchers in this new area, a massive effort called the Human Genome Project has been launched to produce a comprehensive genetic map of the human body. The aim is to locate the approximately 100,000 human genes by the year 2005.

Already, researchers have found that a defective gene responsible for spina bifida and neural tube defects can be neutralized if the mother consumes high levels of folic acid. The malfunctioning gene produces a less efficient enzyme that requires the extra folic acid to do its job. As a result, the FDA has ordered manufacturers to add folic acid to most products made from grains, such as bread, flour, and pasta. Researchers also suspect that the same treatment on adults with the aberrant gene could stave off heart disease.[12]

In the mid-1990s, the BRCA-1 gene, believed to be a factor in almost all breast cancers, was isolated. When functioning properly, it is believed to fight cancer, but in most women with breast cancer, there is a defective or malfunctioning protein attached to the gene. The discovery could lead to genetic alterations to prevent women from developing breast cancer.[13]

Pharmaceutical companies are also undertaking extensive research to find drugs that will interfere with certain genetic activities. To alleviate the effects of osteoporosis, a bone disease, a drug could possibly be designed to block signals to cells that are constantly destroying bone. Normally the work of these cells is countered by other bone-building cells. However, in older women who lack sufficient calcium, these bone-building cells work more slowly than their destructive

11. Gina Kolata, "In the Rush Toward Gene Therapy, Some See a High Risk of Failure," *New York Times,* July 25, 1995.
12. "Flour Will Be Fortified to Fight Birth Defects," *The Record,* March 1, 1996.
13. Gina Kolata, "Research Links Single Gene to Almost All Breast Cancers," *New York Times,* November 3, 1995.

counterparts. According to James Vincent, chairman of Biogen, Inc., "The impact of gene therapy on the pharmaceutical
industry has the potential to equal, and quite possibly to sur-
pass, the advances in diagnosis and treatment that the
biotechnology revolution has brought about in its first 25
years."[14]

Scientists also believe they may have found genes
linked to manic-depression, schizophrenia, impulsive traits,
Alzheimer's disease, Lou Gehrig's disease, and scores of other
afflictions. Already, drugs that affect the brain's chemistry
help more than three out of five mental patients. And it
seems highly possible, researchers estimate, that Alzheimer's
and Parkinson's, the two most common brain disorders, may
be brought to rein early in the twenty-first century.

Another encouraging development is telemedicine, which
allows specialists, using two-way television, electronic stetho-
scopes, and long-distance x-ray transmission, to render
diagnoses on patients hundreds of miles away.[15] Similarly, a
device developed to detect breathing and heart beats on the
battlefield could be mounted above a baby's crib to sound the
alarm against the sudden infant death syndrome.[16]

Recently, a research team at the Max Planck Institute of
Biochemistry near Munich developed a technology that permits, remarkably, the neurons of leeches to exchange information with silicon chips. In coming years, such technology could
possibly be used to create artificial limbs that obey orders
from the brain.[17]

Better nutrition, discussed in Chapter Two, also plays a
key role in the improving health of people around the world.

14. Lawrence M. Fisher, "Bristol-Myers and Biogen to Invest in Gene Therapy,"
New York Times, n.d.

15. Bill Richards, "Hold the Phone," *Wall Street Journal*, January 7, 1996.

16. Emory Thomas Jr., "Vital Signs Can Be Measured Without Human Touch,"
Wall Street Journal, February 7, 1996.

17. "Could This Be the First Step to a Plug-In Brain?" *New York Times*, August
27, 1995.

In the 1950s, Americans consumed an average of 3,100 calories each day, compared with some 3,700 calories daily in recent years, *SAUS* statistics show. In the same period, average daily cholesterol consumption has dropped from 510 milligrams to 410. Thus, though people are eating more food, they are eating in a healthier fashion.

Vitamin and mineral supplements that producers put into food are also making people healthier. For example, iodized salt has virtually eliminated the once-common goiter, a disfigurement of the neck caused by a disease of the thyroid gland.

Looking ahead, the U.S. Census Bureau projects that average life expectancy at birth will reach 82.6 years by 2050, which breaks down to 85.6 years for women and 79.7 for men. Looking still further into the future, a research team in Denmark maintains that children born today in America will have an average life span of 100 years. "It will be 80 years before they are 80," says researcher James Vaupel. "In those years there will be a lot of health and biomedical progress."[18] *Panati's Extraordinary Endings of Practically Everything and Everybody* estimates that "the best scientific estimate [is] 115 years for the time that a human being could live if he or she were spared all the life-shortening diseases and infections."[19]

If healthy citizens are indeed a country's greatest asset, as Winston Churchill claimed, the trends now evident bode exceedingly well for all nations as a new century approaches.

18. Jill Smolowe, "Older, Longer," *Time* Special Issue, "The Frontiers of Medicine," Fall 1996.

19. Charles Panati, *Panati's Extraordinary Endings of Practically Everything and Everybody* (New York: HarperCollins, 1989).

4.

Working Conditions

*"Which of us . . . is to do the hard and dirty work for the rest —
and for what pay? Who is to do the pleasant and clean work,
and for what pay?"*
—John Ruskin (1819–1900), lecture

- Corporate downsizing affects only 3 percent of the total U.S. labor force.
- Self-employed Americans earn 40 percent more per hour than people who work for others.
- The number of U.S. telecommuters is expected to exceed 15 million by the start of the next century.
- Fringe benefits now equal more than 18 percent of total compensation.

The Changing Nature of Work

The nature of work tends increasingly to be "pleasant and clean," as opposed to "hard and dirty." For more and more people in more and more parts of the world, the workplace has become a remarkably agreeable place to be. Work that is dull, distasteful, and often dangerous is performed increasingly by machines, freeing people to undertake more challenging and more satisfying tasks. This process of automation, moreover, helps spur labor productivity, which in turn provides, as we witnessed in the Introduction, a mighty lift to living standards.

Among the many benefits that flow from this transformation of the workplace is a sharp decline in the number of deaths that occur on the job. The annual total in the United

States, for instance, now stands at about 5,000, down from 14,100 as recently as 1965. With a rapidly expanding U.S. work force, the decline in the *rate* of on-the-job deaths is sharper still. It stands at about 4 per 100,000 workers, down from 20 per 100,000 in 1965. On the decline as well are disabling workplace injuries, though data are less precise than for deaths. In sum, today's workplace is spectacularly safer than years ago.

On-the-job mishaps have always tended to run high in such endeavors as farm work, coal mining, and railroading. But these occupations constitute sharply reduced shares of the overall labor force throughout the industrial world. Farm workers, for instance, are likelier than most to receive accidental cuts that can easily become infected around animals and soil. However, thanks largely to the automation of farm work, farm workers now make up barely 2 percent of the U.S. labor force, down from about 40 percent early in this century.

Today's much safer workplace also reflects a concerted effort by employers and government to prevent accidents. For example, a program called Maine 200, launched in 1993, encourages employers in that state to identify workplace hazards themselves and take corrective action before injuries occur, rather than wait for the Occupational Safety and Health Administration (OSHA) to blow the whistle. OSHA selected two hundred Maine companies with the highest number of work-injury claims to participate in the program. These firms comprised only about 1 percent of the state's employers, but accounted for some 45 percent of its workplace injuries, illnesses, and fatalities. OSHA offered each of these companies a choice: to identify and correct workplace hazards themselves and implement corrective measures or face increased OSHA inspection and supervision. Nearly all chose the first alternative, with stunning results: In the eight years before the program was launched, OSHA identified some 37,000 workplace hazards at 1,316 work sites. In the first two years under Maine 200, the employers themselves identified as many as 174,331 workplace hazards! They had managed, moreover, to correct 118,671 of them.

A single participant in Maine 200—Great Northern Paper Co.—was able to identify nearly 30,000 workplace hazards by using teams of employees to inspect "every inch of our vast facilities," according to Glenn L. Rondeau, Great Northern's manager of safety services. All of these hazards have been corrected, or soon will be, at a cost of some $32 million.[1]

For companies unwilling to make such efforts, the penalties extend far beyond the loss of injured personnel. In 1995, for example, OSHA fined a southwest Missouri company $100,000 for allowing two employees to work on a scaffold without any protection against falling. In another case, a small excavating company was fined $168,000 for violating OSHA excavating rules.[2]

When OSHA determines that a safety or health violation is "egregious," much larger fines are often levied. The agency fined a construction firm in Guam $8.3 million in 1995 after a worker plunged fifty feet to his death. There were no railings or harness ropes, so that workers were walking across steel beams with no protection, OSHA reported.[3] There is no doubt that fines of such magnitude help promote a safer workplace.

Another key factor in the safer workplace is the remarkable growth of service-type jobs, which reflects that sector's expanding role on the economic stage. Before 1970, Americans consistently spent more on nondurable goods, such as food, clothing, heating oil, and gasoline, than on services. But in recent years spending on services has exceeded combined spending on both nondurable goods and durable goods, such as motor vehicles and household appliances.

1. Mary Anne Lagasse, "Great Northern Safety Efforts Catch Federal Eye," *Bangor Daily News*, October 18, 1995.
2. Clarissa A. French, "OSHA Uses Fines to Force Compliance," *Springfield (Missouri) Business Journal*, October 9, 1995.
3. Rochelle Sharpe, "Work Week," *Wall Street Journal*, October 31, 1995.

This shift in spending patterns is mirrored in the labor force. In the early 1950s, the U.S. economy's goods-producing sectors provided jobs for four of every ten payroll workers. Recently, however, this ratio has fallen to nearly two in ten, a record low, while service jobs have proliferated. The table below traces this global swing to service work.

This rise in service jobs reflects in large measure the ability of factories and farms to turn out more goods with fewer people, which has brought a sharp increase in productivity. Early in this century, for example, railroad track and ties were laid manually, and coal and grain were shoveled by hand into railcars and the holds of ships. Such strenuous, dangerous, and time-consuming work is now done swiftly and much more safely by machines.

Today, service jobs are sprouting even at companies that are essentially manufacturing concerns. Production-line employees still perform crucial work at auto factories, for instance, but now each such worker is backed up by an expand-

Percent of Civilian Employment in Service Jobs

	1970	1990
Australia	57	70
Britain	54	70
Canada	63	72
France	48	65
Germany	43	58
Italy	40	59
Japan	47	59
Netherlands	56	70
Sweden	54	68
United States	62	72

SOURCE: U.S. Dept. of Labor

ing staff of service workers who, using computers, keep a close tab on inventory and sales levels, as well as other matters important to achieving more efficient, steadier production runs.

Many service jobs, I should note, are well-nigh impervious to automation. How do you mechanize the work performed by a barber or a baby sitter? Not easily. Machinery's inability to replace human effort in many services makes productivity gains more difficult to achieve in the service sector. And it helps explain the spread of service-sector jobs: Even in a time of spreading automation, they are hard to replace. No wonder the U.S. Labor Department projects a steep 69 percent increase in the number of manicurist jobs, for instance, in the next ten years. Who wants to trust a machine to clip one's nails?[4]

In many ways, this swing to services is reminiscent of an earlier transition in the developed countries: the shift from agricultural to industrial economies.

A More Stable Workplace

The spread of service jobs injects an element of stability, as well as safety, into the workplace. When a nation's economy enters a recession, unemployment invariably rises, largely because manufacturers must work down excessive inventories as demand weakens. In turn, this forces layoffs and even plant closings. But services, by definition, can't be stored. So the service sector is largely free of the inventory cutbacks that still bedevil manufacturers when recessions strike, even with the increased use of computers.

With the growth of service jobs around the globe, unemployment now often peaks in recessions at lower levels. In

4. Glenn Burkins, "Work Week," *Wall Street Journal*, March 26, 1996.

four of the five U.S. recessions between 1969 and 1991, for example, the jobless rate remained below 9 percent of the labor
force. When the recession of 1990–91 was near its low point,
the rate never reached even 7 percent of the labor force. For
perspective, in the depressed 1930s, when service jobs played
a smaller role than nowadays, the unemployment rate was in
the double-digit range for prolonged stretches and briefly
reached an excruciating 25 percent.

For all of this increased stability in the workplace, the
perception remains widespread that job security has deterio-
rated greatly in recent years. As companies continue to
"downsize" and "restructure" in the name of becoming more
competitive, the argument goes, layoffs increasingly are the
rule. But if job security were deteriorating, one would expect
workers to be spending less time at one company. And this
isn't the case. The percentage of American workers who have
remained with the same company for at least a decade has
changed little. Since 1973, in fact, it has *risen* for most workers, as the accompanying bar chart shows.

This pattern hardly squares with the prevailing notion
that job security has diminished. Note in the chart that
tenure for employed females has actually increased within all
age groups.

Overall, the only decline in tenure—and it is very
slight—is among workers between the ages of fifty-five and
sixty-four years. Not shown in the chart is that the share of
these older workers remaining with one company for as long
as twenty years didn't decline at all in the two decades. It
stood at 21 percent in 1973 and was still 21 percent two
decades later.[5]

Victims of corporate downsizing, I should add, represent
only some 3 percent of the total U.S. labor force. "Because

5. Gene Epstein, "The Demise of Job Security in the U.S. Is More Fiction Than
Fact," *Barron's*, April 10, 1995.

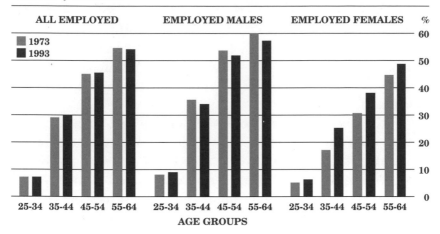

HOLDING GROUND

▶ The percentage of workers across all age groups working for the same company for 10 years or more stayed about the same from '73 to '93. But while the percentage of males on the job for that long dipped somewhat, the percentage of females noticeably increased.

SOURCE: *Barron's,* U.S. Census Bureau

downsizing gets headlines, we tend to write about it as if it's happening to the whole labor force," says Dallas Salisbury, president of the Employee Benefit Research Institute, while "in truth, it's happening to only a sliver."[6]

Meanwhile, the size of the U.S. work force has grown substantially. Headlines naturally focus on employment cutbacks at giant employers, such as AT&T, Sears, and Boeing. As a result, offsetting job gains at hundreds of much smaller firms tend to go unnoticed. Between the spring of 1991, when the 1990–91 recession ended, and the spring of 1996, for instance, the number of civilian workers in the United States rose by about 8 million. This sounds more like corporate *upsizing* than downsizing. Meanwhile, other data show that the layoffs of the 1990s, as a percentage of employment, have been

6. Robert Lewis, "A Surprising Stability," *AARP Magazine,* April 1996.

running close to a fifty-year low![7] The only major downsizing
has been in the governmental sector, which many Americans
regard as in sore need of some belt-tightening.

Particularly heartening is a recent reduction in the time
it takes older executives—those over age fifty—to find new
work after they have been laid off. The average time fell to
only 3.52 months in 1995, down from 3.72 months in 1994.
Such numbers hardly square with much-publicized concern
that high-level people who are laid off late in their careers
face a hopeless job search.[8]

Still other numbers belie the widespread idea that job
growth was far brisker in the "golden" years of the Reagan
presidency than of late. In the first three years of Bill Clinton's presidency, overall employment expanded at an average
annual rate of 2.4 percent, almost as rapid a pace as the 2.6
percent annual gain achieved in the comparable period of the
Reagan presidency. The slightly sharper increase under Reagan, moreover, was due entirely to strong growth in governmental jobs. If one considers only private-sector job growth,
the rate of gain, averaging 2.7 percent a year, is identical in
each period.[9]

Not only has job tenure held steady and employment increased at a healthy pace, but the U.S. work force has grown
vastly more competitive. Again, this is not the common perception. By and large, however, manufacturers have managed
remarkably well to hold down costs and compete with greater
vigor in world markets. This is evident in comparisons of
labor costs per unit of factory output for major industrial nations. So that oranges aren't compared with apples, these
data are expressed in terms of the U.S. dollar's value against

7. H. Eric Heinemann, "The Downsizing Myth," *New York Times*, March 3,
1996.

8. Raju Narisetti, "Work Week," *Wall Street Journal*, March 19, 1996.

9. Rep. Pete Stark (D. Calif.), "Clinton Job Growth Matches Reagan's" in a letter to the *Wall Street Journal*, March 26, 1996.

the various other currencies. In the years 1979–85, such costs in U.S. factories rose at an annual rate of about 4 percent, while they fell in most of thirteen other nations surveyed—a clear sign of declining U.S. competitiveness. The sharpest drop in unit costs, a decline of 9.6 percent annually, occurred in Belgium and only Taiwan showed a faster increase than the U.S. cost rise. Since 1985, however, there has been a remarkable improvement in the U.S. showing. Unit labor costs at American factories have edged down, on average, while comparable costs for other key nations have risen substantially. In 1985–90, for instance, unit labor costs increased at an annual rate of 15.6 percent in Germany, 14.3 percent in Italy, 11 percent in France, 10.8 percent in Britain, and 10.3 percent in Japan.[10] The U.S. performance reminds me of a once-flabby athlete who, to the detriment of competitors, has trimmed down and toned up. The American work force, I should add, is further along than most others in this inevitable, if occasionally painful process.

In this regard, however, it is simply not correct to say, as some analysts do, that the global shift to services has brought a sharp decline in the "quality" of work. The perception is that, as service jobs proliferate, well-paid production-line personnel are being converted into poorly paid hamburger flippers. In fact, as researchers at Syracuse University report, in every year since 1984—except for the 1990–91 recession—more U.S. families have seen their incomes rise than fall. Since 1990, moreover, better-paying jobs have expanded faster than lower-paying ones.[11]

Other data compiled by the U.S. Department of Labor show that in country after country the largest concentration of service jobs—by far—is coming in a category comprising

10. Alfred L. Malabre, Jr., "The Outlook: Protectionist Calls Belie Competitiveness," *Wall Street Journal*, January 27, 1992.

11. David R. Sands, "No Whining Paradise," *Insight*, April 1–8, 1996.

relatively well-paid, challenging service work: jobs in public administration, education, health, and recreation.

In France, there are some 6.6 million service workers in this category, compared to only 3.7 million workers in the category of service work that covers jobs in retail and wholesale trade, hotels, and restaurants, which would include any hamburger flippers. The picture is much the same elsewhere. There are some 8 million British service workers in the relatively well-paying category and only 5.3 million in the lower-paying one. In the United States, similarly, there are about 38 million workers in the better-paying category and 26 million in the worse-paying one.

The better-paying category is not only the largest employment group in the service sector, but the fastest-growing. In the United States, it expanded 72 percent between 1970 and 1990, compared with a rise of 63 percent in the trade-hotel-restaurant category. In Japan, jobs in the better-paying category increased 64 percent in the two decades, while jobs in the latter category rose only 41 percent.

More Challenging, More Interesting Jobs

The modern workplace, to a degree unthinkable several decades ago, is one where workers themselves enjoy a considerable voice in how their jobs, as well as their companies' business, should be managed. The image of the typical worker as a meaningless cog within a vast corporate operation is plain wrong. In reality, today's workers share increasingly in company decision-making, which adds to the challenge of their jobs and makes them more interesting. This is doubly true, of course, if a person happens to be self-employed, which thanks partly to the wonders of modern communications is more and more likely to be the case.

A more challenging, more interesting workplace also
tends to be a more productive one. Consider what has happened, for example, at United Airlines. Most of United's seventy-five thousand employees banded together in 1994 to buy
a majority stake in the Chicago-based carrier. As part of the
arrangement, most consented to large pay cuts in order to
keep the then-troubled airline in business. Since this employee buyout, sick time at United has fallen some 20 percent, saving the carrier more than $18 million annually;
grievances filed by pilots are down 75 percent; and the cost of
workers' compensation claims has dropped 30 percent. The
new enthusiasm of United's employee-owners shows up in
many ways. Cash prizes and motor vehicles are awarded
every six months to employees with good attendance records.
Pilots and flight attendants now are more flexible about taking on assignments. When the company ran short of pilots for
a while in 1995, United's pilot union agreed to let pilots fly
longer hours, rather than exercise their contractual right to
turn down flights. Not surprisingly, productivity at United
has soared since its employees became its majority owners,
and the price of its stock, by no coincidence, has more than
doubled. Moreover, the workplace at United has grown so
popular among airline employees in general that the number
of pilots seeking jobs there exceeds ten thousand, even
though some airlines pay more.[12]

Events at United are part of an international trend. More
and more employees are becoming, to a remarkable extent,
their own bosses. The number of American workers participating in employee stock-ownership plans (ESOPs), for instance, has swelled to more than 10 million, nearly double the
number as recently as 1983. And it is clear that when employees have a larger share in the ownership of companies where

12. David E. Sanger and Steve Lohr, "A Search for Answers to Avoid the Layoffs," *New York Times*, March 9, 1996.

they work, the workplace tends to be happier, more productive and, most importantly, more remunerative.

In this regard, the Employment Policy Foundation, a Washington think tank, reckons that the highest salaries in the years just ahead will come at corporations where managers share decision-making with teams of workers. Companies that have recently adopted such tactics typically see productivity jump 18 percent to 25 percent, which boosts profits and thus leads ultimately to higher pay.[13]

The work effort put forth at companies like United may suggest that, in the drive to be more productive, these employee-owners are overworking themselves, leaving precious little time for leisurely pursuits. Indeed, this is argued in a recent, widely quoted book, titled *The Overworked American: The Unexpected Decline of Leisure*.[14]

The average American household does put in more time on the job than a couple of decades ago: from the mid-1970s to the late 1980s, the total work-for-pay time put in by families with both spouses employed rose from an average of 83.2 hours a week to 86.2 hours. This increase was offset, however, by a decline in time spent on unpaid work around the home. As a result, leisure time—thanks largely to a proliferation of labor-saving home appliances and more child-care options—was undiminished.[15]

The rising amount of capital that stands behind each job has also helped free people from monotonous, sometimes dangerous tasks to undertake more challenging, more productive work. Investment in the workplace has climbed sharply not only in such developed nations as the United States, Japan, and those in western Europe, but in many developing regions, as the accompanying bar chart shows. Based on World Bank

13. Glenn Burkins, "Work Week," *Wall Street Journal*, March 12, 1996.

14. Juliet B. Schor, *The Overworked American: The Unexpected Decline of Leisure* (New York: Basic Books, New York, 1992).

15. Kristin Roberts and Peter Rupert, "The Myth of the Overworked American," *Economic Commentary*, Cleveland Federal Reserve Bank, January 15, 1995.

data, it tracks the annual growth of capital invested per
worker, as well as GDP per worker, a gauge of productivity
gains. It covers five developing regions during a recent thirty-
year period.

The transition to a more challenging workplace extends
even to such unlikely occupations as the manufacture of
potato chips. In 1990, there were thirty-eight managers at the
Lubbock, Texas potato-chip plant of PepsiCo Inc.'s Frito-Lay
unit, and work on the heavily supervised production line was
repetitive and exceeding boring. But recently the number of
managers was cut to thirteen and teams of hourly workers

Growth rates of GDP and capital per worker.
Data are annual averages for 1960–1990.

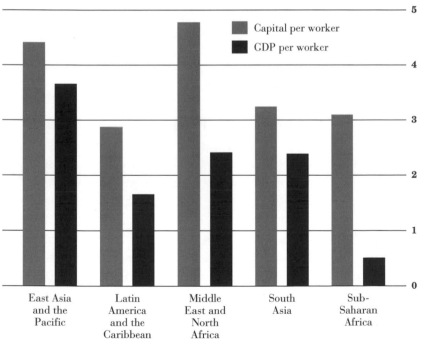

*Capital per worker has risen rapidly in all regions, while output growth
has been slower.*

SOURCE: ILO 1986 with ILO updates; Nehru and Dhareshwar 1991; World Bank data

were organized with authority to decide daily about such matters as quality control and work force levels. In the process,
unit labor costs dropped and productivity jumped, even
though forty hourly employees were added to the payroll.

The team concept "kind of frightened me at first," one
veteran of Frito-Lay's Lubbock production line told an inter-
viewer. "I thought, 'I'm not going to be able to decide anything,' " she recalled. But now she enjoys the team approach
and increased responsibility. "It gives me a sense of pride. I
know my work and what we need to do."[16]

A similar enthusiasm is common among people going
into business for themselves. These entrepreneurs, of course,
must make all the important decisions themselves, as well as
bear the responsibilities. The growing appeal of self-employment is evident in the fact that more than 700,000 new businesses are launched yearly in the United States, a record
pace. The risks are high, but so are the potential rewards: the
self-employed on average earn some 40 percent more per
hour than people who work for others.

Making the Workplace More Pleasant

For millions of workers around the world, a most unpleasant
part of the job has long been the time-consuming, often ardu-
ous task of simply getting back and forth between home and
place of employment each day. However, thanks to the wonders of modern communications, more and more workers are
becoming *telecommuters,* people who mainly work at home,
but still manage to keep in close touch with company headquarters by way of telecommunications.

Close to 9 million Americans now telecommute to and
from work each day. Indeed, these stay-at-home employees

---

16. Wendy Zellner, "Team Player," *Business Week*, October 17, 1994.

represent the swiftest-growing portion of the U.S. work force. Link Resources, a market research firm in New York, projects that the number of U.S. telecommuters will pass 13 million by 1998 and exceed 15 million by the start of the next century.

Telecommuting also helps employers trim overhead by reducing the amount of office space needed. In Washington, D.C., the General Services Administration, the federal government's official landlord, estimates that the government could save as much as $2 billion by the end of this decade if one-third of some 750,000 federal employees worked closer to home or at home. Recently, only some 3,000 federal employees telecommuted, but plans call for as many as 60,000 federal telecommuters by 1998.

Meanwhile, some two-thirds of the largest 1,000 U.S. corporations have established telecommuting programs in the last few years. Some 90 percent of these companies, moreover, report that the moves have helped them reduce costs, increase productivity, and, most importantly, improve employee morale. "The cost of supporting a worker in a cubicle downtown office is very high and getting higher," according to Richard Cooper, an official of Yellow Freight Systems, a Kansas-based trucking company. "It's a major cost that companies can reduce or eliminate, in many cases, by having the employees work from home."[17]

Telecommuting not only makes the job cheaper for employers and more pleasant for employees, but serves to spur productivity. A worker who spends as little as 20 minutes driving to the office each day still spends more than 160 hours a year in the car just to commute. That's roughly four weeks of work! Many commuters, of course, spend much more than 20 minutes driving to work, as suburban sprawl extends farther and farther from downtown office areas.

17. Scott Sutell, "Companies Catching on to Telecommuting," *Crain's Cleveland Business*, October 30, 1995.

The home, to be sure, isn't always the ideal workplace. Some people don't have enough work space there. Many have children at home during working hours, especially when school is out, which can be distracting. Some people feel isolated from colleagues and miss proverbial gatherings around the water cooler and at the bulletin board. For such people, however, suburban "telework centers" are springing up. These centers provide work space away from home—but not as far away as downtown—where people who can't easily work at home still can work *near* home.

One such telecommuter is Martin M. Gertel, a federal auditor. Formerly, he faced a 90-minute trip each way between his suburban home and his office in the District of Columbia. Now he works at a telework center only 20 minutes from his home. Among other benefits, the shorter trip allows him to see much more of his two young daughters. Previously, they were usually asleep by the time he got home at night.[18]

Whether people work at home, near home, or at a distant office, the modern workplace is more pleasant in that it provides far more so-called fringe benefits for employees than years ago. Over 97 percent of people working full-time at private U.S. businesses with at least 100 employees receive paid vacations, 82 percent get paid medical care, 91 percent receive paid life insurance, and 78 percent are covered by paid retirement plans. Other widely held benefits, paid partly or entirely by employers, include parking space, educational aid, child-care and elder-care assistance, recreational facilities, and treatment for drug and alcohol abuse.

In all, these benefits constitute a swelling share of the compensation that U.S. workers receive from their employers. Indeed, the percentage seems far too large for a pay item still

18. Kirstin Downey Grimsley, "GSA Pilot Program Introduces Federal Workers to Telecommuting," *Washington Post*, October 26, 1995.

Increase in Employee Benefits 1950–1995

	Total Pay in Billions	Percent Pay in Fringe Benefits
1995	$4,209.1	18.8
1990	3,297.6	16.8
1980	1,644.4	16.3
1970	603.9	10.3
1960	294.2	8.0
1950	154.6	5.0

SOURCE: U.S. Department of Commerce

referred to as merely "fringe" benefits. Its growth is traced in the accompanying table.

Pacific Mutual Life Insurance Co. in California is among the many employers stepping up these non-pay benefits. Among other things, it provides maternity leave of up to twelve weeks and, after the child arrives, it offers a child-care program that allows up to $5,000 annually in reimbursable expenses. The company also provides nursing stations for mothers with infants and will reimburse employees for up to half of various educational costs.

Finally, the modern workplace is more agreeable to the extent that it harbors far less discrimination than years ago. Blacks and women, particularly, are treated much more fairly nowadays than, say, in the early post-World War II years. Laws are now in place that ban discriminatory employment practices and impose stiff penalties on employers who disregard them.

In sum, working conditions have vastly improved in recent decades, to a point where today's workers are far better off, by and large, than workers have ever been before.

5.

▬▬▬▬

Technology

"One of the ironies of the 20th Century is that Marxist theorists, as well as their critics, such as George Orwell, correctly noted that technological developments can profoundly shape societies and governments, but both groups misconstrued how. Technological and economic change have for the most part proved to be pluralizing forces conducive to the formation of free markets rather than repressive forces enhancing centralized power."
—Joseph Nye and William Owens, *Foreign Affairs*,
March–April 1996

- Every dollar spent on space R&D generates an additional seven to nine dollars in new economic activity.
- Sales of goods and services derived from satellite technology will soon reach $54 billion annually.
- The Electronic Funds Transfer Expansion Act will eliminate the governmental issue of paper checks for every government payment other than tax refunds.
- Child-proof medicine bottles have halved the number of childhood deaths due to accidental poisoning.

The Pace Quickens

The technological revolution that surrounds us has, among other benefits, helped greatly to improve our living standards. Killer diseases have been brought under control. New agricultural techniques have enabled us to keep pace with a burgeoning world population. The Internet has given us instantaneous access to a previously unimaginable storehouse of knowledge. Advances in transportation have dramatically shrunk the world in which we live.

The list goes on, as this chapter will show. Advances in technology have brought improvements in virtually every area of human endeavor—from space exploration to energy production and from construction to commerce. What's more, the pace of technological innovation is quickening. In the 1960s, the average lag was thirty years between a basic discovery and its practical application. Today, discoveries become obsolete in a fraction of that time.

"I have been through eight or 10 technology shifts," says Ray Emery, vice president for worldwide operations at Microsoft. Recalling his twenty years in the auto industry, he

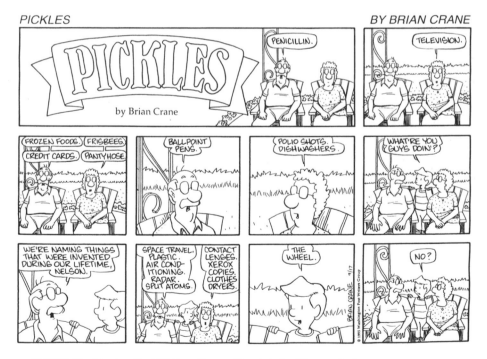

PICKLES   BY BRIAN CRANE

© 1995 *Washington Post* Writers Group. Reprinted with permission.

says that when the shift was made to aluminum engines "it obsoleted the iron industry, but once we got through it we had a lighter, better, more efficient vehicle."[1] When aluminum first came into use, it was so expensive to produce that it was used mainly for jewelry. But with technological advances, its production cost has declined sharply and its use, by no coincidence, has greatly expanded.

A huge rise in research and development (R&D) expenditures has helped make possible such technological gains. In 1960, for example, the United States spent $13.5 billion on R&D, of which 55 percent was defense and space related. Recently, by comparison, annual R&D expenditures totaled $171 billion, of which less than 25 percent was for defense and space. Even adjusted for inflation, the latest spending total represents a 152 percent increase from 1960. Since 1970, moreover, employment of R&D scientists and engineers in the United States has increased nearly 77 percent, to more than 960,000, according to the *SAUS*.

Major advances in research are being achieved through "digital science," which *Business Week* calls "the most fundamental change in scientific methodology since Isaac Newton laid the foundations 350 years ago." Right now, computers can capture the activity of proteins during only 1.5 billionths of a second and show what happens in enough detail to take up a seven-hour video. Tomorrow's computers are expected to be able to calculate how electrons, protons, and neutrons may be reassembled to create a new drug or metal, for example. Already, major automobile companies, DuPont, and Caterpillar are using digital science to improve their products.[2]

1. Judith Berck, "Microsoft Issues Pink Slips To Most 3.5-Inch Floppies," *New York Times*, March 11, 1996.
2. Russell Mitchell and Otis Port, "Fantastic Journeys in Virtual Labs," *Business Week*, September 19, 1994.

A Three-Dimensional Molecule

Three times this century, entire industries have blossomed from the invention of a new molecule. The inventions led to nylon, polyethylene, and Plexiglas. Now, an even more important molecular breakthrough may be in the offing. Scientist Donald Tomalia has created the dendrimer, a three-dimensional molecule. Its structure could make skin creams and perfumes more potent, allow computer chips to store far more data, and help develop electronic components the size of a single molecule. It also could be a means of carrying genes into the body to treat heart disease and cancer.[3]

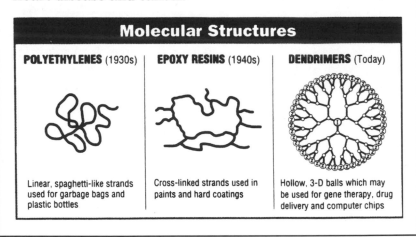

Molecular Structures

POLYETHYLENES (1930s)

Linear, spaghetti-like strands used for garbage bags and plastic bottles

EPOXY RESINS (1940s)

Cross-linked strands used in paints and hard coatings

DENDRIMERS (Today)

Hollow, 3-D balls which may be used for gene therapy, drug delivery and computer chips

The Benefits of Space Exploration

Some of today's most startling technological advances have come in space exploration. In 1928, American inventor Nikola

3. Ama Camber Na, "Persistent Inventor Markets a Molecule," *Wall Street Journal,* April 22, 1995.

Tesla stated: "No rocket will reach the moon save by a miraculous discovery of an explosive far more energetic than any known. And even if the requisite fuel were produced, it would still have to be shown that the rocket machine would operate at 459 degrees below zero—the temperature of interplanetary space."[4] Less than thirty years later, Sputnik was spacebound, and by 1969 men were walking on the moon. Since 1957, nations of the world have successfully completed more than 3,733 space launches!

Film critic Roger Ebert recently recalled that early in the movie *Apollo 13* (Apollo 11 landed men on the moon) as astronaut Jim Lovell takes the press on a tour of the Kennedy Space Center he notes that the center has a computer "that fits into one room and can send out millions of instructions—and I'm thinking to myself, I'm writing this review on a better computer than the one that got us to the moon."[5]

Since Apollo 13 was launched, advanced technology has allowed space probes to be launched toward the outer limits of the solar system. Recently, Pathfinder made a successful landing on Mars, and dispatched the Sojourner robot to collect information on the Red Planet, including possible signs of interplanetary life.

James A. Lovell, Jr., commander of the Apollo 13 lunar mission, recently speculated about "a few of the things that would happen if we turned off the hundreds of satellites orbiting our globe." A back-packer lost in a blizzard can't be located without help from the Global Positioning System transmitter. A tornado injures an elderly woman because she received no satellite warning from the National Weather service, and she dies because a neurologist can't assist the local doctor via telemedicine. A woman's new job is jeopardized

4. David Wallechinsky, "Bad Predictions," *Parade*, September 10, 1995.
5. Roger Ebert, "America's Derring-do Resurrected," *The Record*, June 30, 1995.

Apollo 11 commander Neil Armstrong took this photo of astronaut Edwin M. Aldrin walking near the Lunar Module on July 20, 1969.

because a car dealer can't run an instant credit check, leaving her stranded.[6]

Lovell estimates that sales of goods and services derived from satellite technology will soon reach $54 billion annually and estimates that more than thirty thousand commercial products have already evolved from space technology, with thousands more in the pipeline. At present, every dollar spent on space R&D generates an additional seven to nine dollars in new economic activity, he reckons, adding that "this is merely the initial trickle of benefits. Space has the potential to improve immeasurably the lives of everyone on earth."[7]

The Ballistic Missile Defense Organization of the U.S. Department of Defense provides one example of the way in which space research is transferred to civilian use. So-called adaptive optics, used to eliminate atmospheric "ripple" during the sighting of satellites, is now also being used by ophthalmologists to diagnose retinal and other eye diseases. Additionally, the Baylor Research Institute is experimenting with light-activated chemical dyes that can "bond" collagen during corneal surgery.[8]

Electrifying Advances

Energy consumption in the United States has increased more than one hundred times in the past century. Electric lights have replaced candles and oil lamps. Some sixty years ago, people placed cards in their windows to show how many pounds of ice they wanted delivered. The ice blocks, carved

6. James A. Lovell, Jr., "Life After Apollo 13," *Wall Street Journal*, March 22, 1996.

7. Ibid.

8. "Going Ballistic," *Wall Street Journal*, June 6, 1996.

out of frozen lakes in the winter and stored in straw, were used to chill food in metal-lined, wooden ice-boxes.

Today, frost-free refrigerators automatically churn out ice cubes for drinks and keep food both chilled and frozen. Oil, gas, and electricity have replaced coal and wood in heating homes. In the last century, the whaling industry flourished by furnishing oil to light lamps. Today, conservationists are fighting to save what remains of these once bountiful mammals.

In 1940, according to the *SAUS,* the United States produced 180 billion kilowatt hours of electricity. Since then, yearly electricity production has increased to more than 3 *trillion* kilowatt hours, a gain of more than 1,500 percent! Meanwhile, in just the last quarter of a century, world energy consumption has risen by 75 percent, with electricity consumption increasing more than six-fold.

A major technological shift is the increasing use of renewable energy sources, such as solar power, hydroelectric power, and wind. There has been a huge increase in the use of nuclear power as well, despite recurring worry over its safety.

The nuclear powered generator came on stream in 1956. Fourteen years later, there were 18 nuclear power plants in the United States alone, generating 21.8 billion kilowatt hours of electricity. By 1995, there were 109 plants in the United States, generating 673 billion kilowatt hours—a thirty-eight-fold increase in nuclear-generated energy! Today, nuclear generation accounts for more than 22 percent of all electricity in the United States, up from 1.4 percent in 1970. And worldwide, the number of nuclear reactors has risen to 421 from 64 in 1970. The resulting gain in electricity generated has been dramatic—from 74 billion kilowatt hours to more than 2.1 trillion, nearly a thirty-fold increase.

In 1970, energy from nonfossil fuel sources accounted for less than 3 percent of all sources of heat and power. Today, with consumption up more than one-quarter, these energy sources contribute nearly 10 percent, according to the *SAUS.*

The U.S. Department of Energy forecasts a 60 percent increase in energy consumption worldwide in the next twenty years, spurred by a sharp increase in demand in such major developing countries as China and India. To meet its growing need for energy, India is already experimenting with wind and solar power generation. Likewise, many developing nations in arid lands with plenty of sun are investigating the use of solar power plants like one that Southern California Edison opened in 1996 in the Mojave Desert. The plant employs a technology that allows the sun's heat to be stored in molten salt for use as needed to generate electricity on demand.

Construction

In many American cities, the word ferry brings to mind a time when traveling across the river was a matter of navigation

Solar power plant, Southern California Edison, Mojave Desert.

over water rather than driving over a bridge. To be sure, bridges have been around for centuries. But only 180 years ago, the longest bridge in the world reached out just over 400 feet across the Schuylkill Falls near Philadelphia.[9] And 30 years ago the Verrazano Narrows Bridge in New York claimed the record, spanning 4,260 feet, a ten-fold increase in 150 years.

Admittedly, the Verrazano Bridge is a single-span structure. However, when the Akashi Kaikyo Bridge in Japan is completed, probably in 1998, it will have but one span that will be 50 percent longer than the Verrazano's. As amazing as this seems, today's longest road and rail bridge is more than ten times as long as the Verrazano, forming an 8.2 mile link between Japan's main island of Honshu and the island of Shikoku.[10]

Current suspension bridge technology places a theoretical limit of around 10,000 feet on single-span bridges. Longer spans, it is believed, would collapse of their own weight unless current steel cables can be replaced with synthetic material now under development. But even without new materials, designers believe that longer distances can be traversed by integrating the cable-stayed technology used on bridges up to around 3,000 feet with new suspension technology. T. Y. Lin, a San Francisco bridge designer, has proposed building such a bridge across the Strait of Gibraltar, between Spain and Morocco. It would be 8.7 miles long and would include two main spans of more than 15,000 feet each.[11]

Even that length pales in comparison with the Channel Tunnel, which was completed in 1994. The "Chunnel" runs thirty-one miles beneath the English Channel, realizing the long-time dream of a "land" link between France and England.

9. Norris McWhirter and Ross McWhirter, eds., *The Guinness Book of Records* 1976 (New York: Bantam Books, 1976).

10. Peter Matthews, ed., *The Guinness Book of Records* 1996 (New York: Bantam Books, 1996).

11. John Pierson, "Bridge Designs Go to Improbable Lengths," *Wall Street Journal,* February 22, 1996.

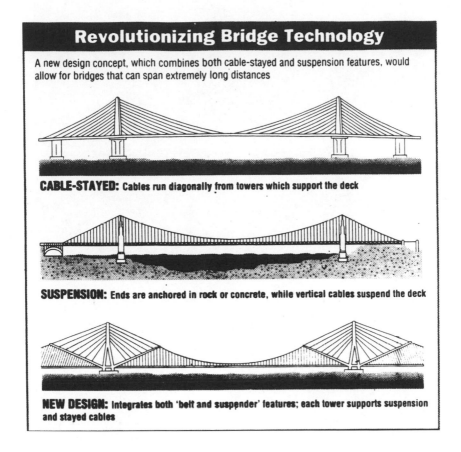

Revolutionizing Bridge Technology

A new design concept, which combines both cable-stayed and suspension features, would allow for bridges that can span extremely long distances

CABLE-STAYED: Cables run diagonally from towers which support the deck

SUSPENSION: Ends are anchored in rock or concrete, while vertical cables suspend the deck

NEW DESIGN: Integrates both 'belt and suspender' features; each tower supports suspension and stayed cables

Improved construction techniques have also made vast differences in city skylines around the world. In 1894, architect Thomas F. Schneider was among the first to use a steel skeleton in constructing the Cairo, a luxury hotel in Washington, D.C. "towering" 165 feet above the landscape. Frightened neighbors on each side of the construction site temporarily moved away in fear of their lives. This "high-rise" hotel created such a storm that in 1910 the city imposed height regulations, and it remains today the tallest private building in the nation's capital.

While Washington took the low road, architects elsewhere around the world continued to look skyward. In 1930,

The world's largest towers, the Petronas Towers, in Kuala Lumpur, Malaysia. The headquarters of the state-owned oil monopoly Petronas is 10 meters (33 feet) higher than the previous record holder, the Sears Tower in Chicago.

the Chrysler Building in New York breached the 1,000 foot
ceiling for the first time. A year later, the Empire State Building, one of the seven modern wonders of the world, was completed, built on a steel skeleton weighing sixty thousand tons.
It rose 1,250 feet, dwarfing its neighbor eight blocks uptown.
Indeed, it towered over all other office buildings for more
than forty years. In the 1970s, however, the Empire State
Building was eclipsed by the World Trade Center towers in
lower Manhattan, and then by the Sears Tower in Chicago. In
1988, there were eight office buildings more than 1,000 feet
tall, all of them located in the United States. In 1996,
Malaysia completed two towers extending skyward 1,483
feet, a new record. By the end of the century there will be
twenty office buildings that are more than 1,000 feet tall, half
of them in Asia and constructed mainly with the help of
American architectural technology. Even more amazing, architects believe they now have the technology to more than
double the height of today's tallest buildings.

Here is yet another perspective on how high-rise technology is spreading around the world: In 1953 Moscow State
University completed the tallest building outside the United
States. It continued to dominate the Eastern Hemisphere
skyline for more than a decade, but now it has fallen to
twenty-sixth place![12]

A Boon for Business

New software technology is helping manufacturing companies speed up deliveries and squeeze more production out of
existing equipment. New programs can simulate the actual
production and supply process in detail, highlighting each operation's specific limitations. According to Sanjiv Sidhu,

12. *World Almanac 1997* (Mahwah, N.J.: World Almanac Books, 1997).

chairman of i2 Technologies, reductions in the cost of computing power make it possible to represent digitally the complex flow of a production process and calculate the domino effect of any change in that process.[13]

Timken Steel Co. has used the new production technology to boost capacity by 15 percent on existing plant and equipment. Furniture maker Herman Miller, guided by the new software, is able to cut delivery time on orders by 20 percent—again, without adding capacity.[14]

The New Money

Money in various forms has been around for centuries. Checks were being written more than 200 years ago. Basically, little changed in the banking industry for more than 150 years. For those of us growing up in the first half of this century, everything had to be paid for by cash or check.

No more. In 1950, Diners Club introduced the first credit card, which has gone on to generate a multi-billion dollar industry. In fact, the much-vaunted cashless society seems to be on its way. Most large U.S. companies offer employees the option of electronic transfers instead of paychecks. Moreover, a recently enacted federal law, the Electronic Funds Transfer Expansion Act, requires that by the end of 1998 the government use this method to pay all its salaries, pensions, veterans', and social security benefits. Indeed, the law eliminates the issue of paper checks for every government payment other than tax refunds.[15]

Many stores accept debit cards that transfer the necessary funds from the customer's account to the merchant's account

13. Amal Kumar Naj, "Manufacturing Gets a New Craze From Software: Speed," *Wall Street Journal*, August 13, 1996.
14. Ibid.
15. National Association of Federal Credit Unions flyer.

with a single swipe. Plastic debit and credit cards also give their holders instant access to real cash at millions of locations around the world through automated teller machines (ATMs).

Technological advances are expected to bring further major changes to the banking industry's products and delivery systems. Technology "is like a tidal wave," says Hugh L. Mc-Coll, chairman of NationsBank Corp. "If you fail in the game, you're going to be dead."[16] Already, about 1 percent of the U.S. population banks through personal computers while banking by telephone is widespread. NationsBank, for instance, handled some 115 million calls in 1996 at its phone banking centers. Other technological options include interactive television, a kind of in-home ATM screen, and "smart cards" that encode everything from cash balances to medical records.[17]

Where tokens or coins were once required to ride the subway in New York, riders may now use cards encoded with prepaid amounts that can be inserted in the turnstiles, so the fare can be deducted. In Washington, D.C., where the fares vary according to the time of day and the distance traveled, cards are inserted at both ends of the trip, and the appropriate fare is charged.

Oklahoma City and Denver are installing digital parking meters, some of which can take debit "keys" as well as coins. And several communities are testing a Pennsylvania company's motion-sensitive Intelligent Parking Meters that wipe off any remaining time as soon as a car departs.[18]

Electronic Technology

Sir Francis Galton, an English scientist born in 1822, was the first to classify fingerprints, which of course are unique to

16. Nikhil Deogun, "High-Tech Plunge," *Wall Street Journal*, July 25, 1996.
17. Ibid.
18. "Business Bulletin," *Wall Street Journal,* March 21, 1996.

every individual. Detectives from Sherlock Holmes onward have used them to link criminals to the crime scene. Today, fingerprint databases have been computerized to trace criminals from state to state and even country to country. And the electronic use of fingerprints is gaining wider use.

California and New York are using electronic fingerprint scans to make sure people do not apply for welfare payments under different names. Disney World scans the hands of its season pass-holders to make sure the real owner is entering the park. Sensar, a firm in Princeton, N.J., is working on an automated teller machine that scans the iris in a user's eye, rather than relying on just a bank card.[19]

Colt Manufacturing Co. is even developing a "smart gun" for police forces that can only be fired by law enforcement officers wearing a ring-mounted transmitter. The .40-caliber semiautomatic pistol uses radio frequency technology that removes a blocking pin from the trigger mechanism. The transmitter has a range of only a few inches, so the technology is expected to save police lives.[20] It is estimated that one in every six police officers killed by a firearm in the United States is shot by an assailant who seized the officer's weapon.[21] Making this safety feature more widely available could also save the lives of hundreds of young people who are killed annually in firearm accidents.[22]

A relatively new technology, radio-frequency identification, makes it possible to collect tolls from cars simply as they drive past a scanner. Drivers don't even have to roll down the window or stop. The so-called EZ Pass is already in use on bridges and highways around New York City. Motorists using

---

19. "When Criminals Have Got Your Number," *New York Times*, June 16, 1996.

20. Gordon Witkin, "Can 'Smart' Guns Save Many Lives?," *U.S. News & World Report*, December 2, 1996.

21. "Hi-Tech Gun Unveiled to Help Protect Police," *The Record*, September 19, 1996.

22. Gordon Witkin, "Can 'Smart' Guns Save Many Lives?"

toll roads in New Jersey will soon be able to carry a device that will allow a scanner to deduct pre-paid tolls as the car drives through a booth without stopping. With falling prices and more sophisticated technology, a scanner may eventually be able to collect tolls from dozens of cars at once as they speed down designated lanes of traffic.[23]

Radio-frequency identification has many uses. Railroads use the technology to keep track of individual railcars as they shuttle around the country. The U.S. Postal Service uses it to trace the flow of mail through the system. An electronic "greeting card" is mailed, and as it is processed and shipped, scanners in post office ceilings pick up its location and are able to spot bottlenecks. Smaller radio-readable devices are being implanted in dogs, cats, and other pets to help owners locate them. By the year 2000, the radio-frequency identification sales are expected to reach $800 million annually, almost four times the current volume, according to Venture Development Corp., a Natick, Massachusetts market research company.[24]

Some of us can still remember the corner grocer hand-picking a customer's order from the shelves and then adding up the prices on a brown paper bag. Over the decades, we moved to self-service and cash registers. A few years ago, President Bush created a stir when he appeared fascinated by supermarket scanners that read a bar code and then automatically ring up the price, but the scanners were already a familiar sight at most American supermarkets. And now there is even a machine being tested that in a single scan can tally an entire grocery cart full of goods. Of course, scanners do more than add up the bill. They identify the best-selling items, reorder from inventory, and even keep track of individual customer's purchasing patterns.

Vending machines are now able to record sales and inventory and transmit the information to a central warehouse. This

23. Laurie J. Flynn, "High-Technology Dog Tags For More Than Just Dogs," *New York Times*, August 12, 1996.
24. Ibid.

knowledge increases the productivity of workers who restock the machines by about one-third. Today's vending machines can even let owners know when they are being vandalized.[25]

Elsewhere, water and electric utilities are also beginning to use transmitters installed in homes and businesses to read meters automatically, eliminating the need for meter readers.

Home and Away

In Washington state, a computer in the home of a Microsoft executive turns off every light at his bedtime, extinguishes any heat in the fireplace, clicks off the infrared heat over the patio, and closes the blinds. At the less expensive end of the scale, RCA offers a remote control unit costing only $59 that can operate television sets and VCRs through an infrared beam and control lights and appliances with a radio signal. This, I would think, is the perfect technology for a couch potato, who would be able to switch channels and pop popcorn while never rising from the sofa.[26]

Huge technological advances have been made in using the telephone as a wide-ranging communications tool. A type of telephone has been developed that can serve as a fire and burglar alarm system, or respond to a personal medical alarm button. When the telephone is activated by the motion of an intruder, a smoke detector, or a person in need of help, it begins to dial a preprogrammed set of up to ten numbers until it contacts someone able to help. It even hangs up on answering machines and proceeds to dial the next number.

It is now possible for lost drivers too stubborn or shy to ask for directions to rely instead on a computerized naviga-

25. G. Pascal Zachary, "Hard Labor," *Wall Street Journal,* June 17, 1996.
26. "For TV, or Even Popcorn," *New York Times*, September 5, 1996.

tion system to get them to their destination. A small dashboard screen displays local street maps, highlights the best route to a desired destination, and gives voice instructions so the driver doesn't have to look away from the road. This remarkable system receives signals from satellites tracking the auto's precise location and then compares them to map data stored on a compact disk in the car's trunk. Hertz Corp. has about 8,000 such units installed in its rental cars in the U.S., and the system is available as an option in certain models of Porsche, Cadillac, BMW, Acura, and Oldsmobile.[27]

A Better Image

Color photography was invented less than ninety years ago, shortly after man had learned to fly. The Land Camera, which begins developing photographs as soon as the shutter is snapped, is less than fifty years old. Now, color reproduction is commonplace in computers, computer printers, copying machines, television, and even newspapers.

Decades ago, to alter the shading of a photograph to be printed in a magazine required a skilled craftsman. Using a special tool under a magnifying lens, he or she would painstakingly remove or insert individually the microscopic dots of the red, blue, yellow, and black that make up color photos. Today, in contrast, the job is done routinely and quickly on a computerized scanner. Moreover, new backgrounds can be easily scanned in, adding a snowscape, for example, to a photograph taken in summer for use later as a Christmas card.

Improvements in computer printers and digital cameras enable today's photographers to use their personal comput-

27. Robyn Meredith, "Ask Directions? Not When His Car's Got All the Answers," *New York Times,* August 25, 1996.

ers, if they wish, to bypass professional developers. Photos
and negatives from standard cameras can be scanned into a
computer and then reprinted and, if need be, transmitted
around the world.

Bill Gates, chairman of Microsoft, views computerized
photography as the "coolest emerging technology" that exists
today. "Digital cameras are getting to the point where it will
be a mainstream scenario to take digital pictures, put them
into an album and mail them around to your friends," he says.
"Color printers are really improving very rapidly."[28]

The world's leading traditional photo manufacturers, at
the same time, are striving to remain competitive with
greatly improved cameras and film. Errors in the use of the
flash can now be readily corrected, for example, and three
sizes of photos may be taken on the same film.[29]

Safer Transportation

Technology is working to make our lives much safer. This
runs the gamut from new designs that make medicine bottles
tamper-proof—an innovation that in thirty years has halved
the number of children dying from accidental medicinal poisoning—to simulated test crashes for cars and other vehicles.
Auto makers now can crash-test cars merely by computer
simulations based on programs originally written to ensure
that a nuclear bomb wouldn't explode if the plane carrying it
were to crash.[30]

Technology used to train military pilots is now being
used to make truck driving safer. The effects of blowouts, high

28. Rick E. Martin, "Microsoft Chief Looks at the Future of Software," *The
Record,* June 17, 1996.

29. Wendy Bounds, "Camera System Is Developed but Not Delivered," *Wall
Street Journal,* August 7, 1996.

30. Russell Mitchell and Otis Port, "Fantastic Journeys in Virtual Labs."

winds, and ice, for instance, can be simulated. North American Van Lines reports that drivers trained with such simulators have 22 percent fewer accidents than other drivers.[31]

Though training drivers to cope with emergency situations is important, the U.S. Department of Transportation estimates that fatigue or drowsiness still causes 15 percent of all accidents where the driver dies. In response, Evaluation Systems of San Diego has recently developed a computerized system to monitor drivers for fatigue. It works like this: After driving for a specified number of hours, the driver pulls off the road and takes a simulated road test on a monitor. Reactions are transmitted by satellite to a central computer that analyzes them and decides whether the driver is okay to continue to drive. If not, the truck can be disabled.[32]

In another approach, researchers at Carnegie Mellon University are equipping trucks with cameras and sensors that can spot signs indicating that the driver should stop for a rest. Telltale signs include jerky steering, weaving, and failure to use rear-view mirrors. In a serious situation, where the driver actually dozes off, an on-board computer can sound a wake-up alarm.[33]

A Sporting Chance

Technology is even helping today's super-athletes to perform better. Between 1992 and 1996, the American Olympic Committee spent $10 million to develop technology for training and equipment, which helped the United States do so well in the recent summer Olympic games. Some $20,000 went, for

---

31. Richard Bierck, "Steering Clear of Danger," *U.S. News & World Report,* August 26, 1996.

32. "A Computerized Overseer for the Truck Drivin' Man," *Business Week,* September 19, 1994.

33. Richard Bierck, "Steering Clear of Danger."

The SuperBike

The SuperBike II, at 16 pounds, is a technological tour de force with a carbon-fiber frame and a knife-like profile that dramatically decreases its wind resistance.

An early cycle
Bikes in the 1930s were made of steel and had wooden rims.

FPG

Tires
Approximately 1/2-inch wide, tires have paper-thin treads and high inflation pressure to reduce road friction.

Wheels
Carbon-fiber disk wheels are more aerodynamically efficient than spoked wheels, as long as there is no crosswind.

Clothing
Woven polyester skin suits are slicker and more breathable than ordinary spandex gear.

Seat
Seat is attached without any exposed bolts, which would increase wind resistance.

Cranks and bearings
Aluminum and carbon-fiber cranks and bearings transmit racer's power to the bike with maximum strength and least weight.

Brakes
There are none. Rider must backpedal to slow the bike.

Pedals
Pedal design is integrated with racer's shoes so they function as one

Handlebars
Swept forward to help racer assume a drag-reducing tuck position.

Helmet
The dimpled titanium skin of the bike helmet both protects the biker's head and reduces wind resistance.

MAVIC

SIMON CLUDBY

example, for a high-tech scale that lets weight-lifters detect when they aren't perfectly balanced as they press hundreds of pounds of weights. A laser tracking system mounted on rifles helped U.S. marksmen sharpen their aim. A carbon-fiber mast and sail made sailboats faster and more maneuverable. Major improvements were also made in the design of such diverse items racing bicycles, track shoes, and poles used for vaulting.[34]

In sum, technological advances have improved almost every aspect of our lives, from the way we cook and shop, to the way we travel and manage our money. One of my favorite examples, however, comes from the work of biologists at the University of Georgia. They have genetically engineered weeds that can contribute to improving the environment by cleaning up toxic waste. The altered weeds absorb mercury particles through their roots, converting them into a comparatively benign substance.[35]

34. Joannie M. Schrof, "The Winning Edge," *U.S. News & World Report,* July 29, 1996.

35. Robert Langreth, "Altered Weeds Eat Mercury Particles in Lab Experiments on Toxic Waste," *Wall Street Journal,* April 16, 1996.

6.

━━━━━

Political Freedom

━━━━━━━━━━━━━━━

"The capitalist preference for law and due process leads naturally enough to the . . . basic institutions of democracy: The rule of law, limited government, separated powers, and the protection of the rights of individuals and minorities."
—Michael Novak, *Freedom Review*, March–April 1996

"I know of no example in time or place of a society that has been marked by a large measure of political freedom, and that has not also used something comparable to a free market to organize the bulk of economic activity."
—Milton Friedman, *Capitalism and Freedom,* 1963

- Of the world's 192 sovereign states, fully 179, or 93 percent, now elect their own legislators.
- Democracies rarely wage war against each other.
- Political freedom is increasing dramatically around the globe.

Freedom's Growth

Only about a century ago there were but a handful of democracies on earth. Much of the planet was occupied by totalitarian states or subjugated colonial regimes. Today, in happy contrast, democracies are flourishing. The Geneva-based Inter-Parliamentary Union says that of the world's 192 sovereign states, fully 179, or 93 percent, now elect their own legislators. And of these, 69 have held multi-party elections for the first time ever in just the past decade!

There are many aspects to political freedom, of course, as the checklist on page 114 makes clear. But according to a

1996 survey of political freedom by Freedom House, a Washington-based non-profit, seventy-six nations were deemed "free" and another sixty-two nations were rated at least "partly free." A decade earlier, by comparison, Freedom House, which is dedicated to strengthening democratic institutions, determined that only fifty-six nations were "free" and a like number were designated as "partly free."

Even by Freedom House's stringent standards, nearly four dozen nations moved into higher categories of freedom in just ten short years. The list of nations advancing to greater degrees of political freedom includes such diverse examples as Nicaragua and the Dominican Republic in the Americas and Mali and Eritrea in Africa. Meanwhile, the number of nations categorized as politically "not free"—this group includes such states as North Korea and Cuba—declined in the decade to fifty-three from fifty-five.

The sweep of democracy through Africa is especially notable. Approximately three-quarters of the several dozen nations in sub-Saharan Africa have recently undergone some sort of political liberalization. At the end of World War II, by comparison, almost all of Africa was ruled by European states. By 1956, however, much of this colonial area had gained political freedom. In 1957, the former British colony of Ghana gained its independence and Kwame Nkrumah, who came out of prison, become its first president. Asked why he hadn't studied economics while attending universities in Great Britain and the United States, he explained that he had expected to spend his life fighting for Ghana's freedom. He hadn't thought he would actually be running the country. He should have studied economics, as it turned out, because his mismanagement of the Ghanian economy eventually pushed the nation to the brink of bankruptcy, setting the stage for a military coup.

In fact, elected governments have tended to come and go in Ghana, as well as in other former African colonies, since it gained its freedom. However, as Adrian Karatnycky, president of Freedom House, says, the "good news is that many free societies are showing signs of increasing durability, as years of democratic rule and tolerance are creating a stronger infrastructure of civil society, especially in the post-Communist countries of Central Europe and in Latin America."[1]

High on the roll call of nations that have recently moved toward greater political freedom are, of course, the post-Communist regimes of Eastern Europe. The population of these European states approximates 310 million. All of these individuals enjoy a significantly higher degree of political freedom—despite occasional setbacks—than was the case a couple of decades ago.

The spread of political freedom, I should add, extends well beyond the post-Communist nations of Eastern Europe and the former Soviet Union. In Latin America, in contrast to the situation a couple of decades ago, every country except Cuba now boasts an elected government. Elsewhere, consider the remarkable transformation that has recently occurred in Mongolia, a vast, isolated nation in central Asia, sandwiched between China and the former Soviet Union, with 2.3 million people and 28 million cattle.

In mid-1996, Mongolian voters ranging from nomadic herdsmen to city dwellers went to the polls to oust the Communist party from power. Known as the Mongolian People's Revolutionary Party, the Communists had run Mongolia as a Soviet puppet state for seven decades. Viewing the election results, Enkhasai Khan, a leader of the successful Democratic Union Coalition, declared that finally "Mongolia is strongly standing on the path of democracy."[2]

Precipitating this electoral victory, of course, was the fall of communism in the former Soviet Union. After the Soviet collapse in 1991, the Mongolian economy contracted by about one-third. Fuel became so scarce that many Mongolians literally froze to death. Inflation was rampant. But now, with reform of

1. Adrian Karatnycky, *Freedom Review,* January–February 1996.
2. "Communists in Mongolia Are Toppled After 70 Years," *New York Times*, July 3, 1996.

How the Freedom Checklist Works

The range of measures used by Freedom House to gauge the extent of political freedom around the world is extensive. In the survey, freedom entails the right of all adults in a nation to vote and compete for public office, and for elected officials to have a decisive vote in public policies. In addition to such political rights, a nation to be considered free must also offer its citizens a high degree of civil liberty to develop individual opinions that may be quite different from state-endorsed views.

On the survey's "checklist" are such matters as: Are the head of state and legislative leaders elected through free elections? Do the people have the right to organize into different political parties of their choosing? Are the people free from domination by the military, foreign powers, and religious hierarchies? Is there a realistic possibility for the political opposition to the elected leadership to increase its support and gain power through elections?

Additional questions include: Are there free and independent media? Is there open political discussion? Is there freedom of assembly and demonstration? Are all citizens equal under law, with access to an independent, nondiscriminatory judiciary? Are there free trade unions? Are there free religious institutions? Are there such personal freedoms as gender equality, property rights, freedom of movement, and freedom to choose one's residence, one's marriage partner and the size of one's family?

the Mongolian central bank and a privatization program under way for government-run businesses, inflation has been cut by more than three-quarters and the economy is beginning to expand. Such has been the change that Mongolia now ranks among the world's entirely "free" nations, as the global "Map of Freedom," compiled by Freedom House, indicates.

With regard to the map on the next page, it is encouraging to note that the portions of the globe designated as free or partly free far exceed those marked as still not free. I need hardly point out that not so many years ago the map's black area would have been vastly larger. All of Russia, for example, would have been so designated, along with other large areas ranging from South Africa to Chile. It is also encouraging that many of the nations deemed not free seem likelier to move toward a greater degree of political freedom than the reverse. Examples include Cuba, where Fidel Castro's time is clearly running out, and Iraq, where the same may be said for Saddam Hussein. Other nations still marked on the map in black, such as Egypt, Burma, and Indonesia, seem likely candidates to adopt, however reluctantly, less restrictive political systems in coming years.

Different Paths, Same Direction

Such is its pull around the world that political freedom is emerging under the most unlikely of circumstances. Consider the case of Chile, a nation of some 14.2 million that Freedom House now ranks as politically free in all respects. In the 1970s and much of the 1980s, Chile was ruled by the dictatorial power of General Augusto Pinochet. Yet, in retrospect, this dictatorship played a critical role in turning the Chilean tide toward democracy after years of leftist mismanagement. It was only through the clout of a military dictatorship that Chile was able to establish a privatized social security

GREENLAND (DEN)

ICELAND

FAEROE ISLANDS (DEN)

NORTHERN IRELAND (UK)
IRE.
UNITED
ISLE OF MAN (UK)
CHANNEL ISLANDS (UK)
NETHERLANDS
BEL.
LUXEMBOURG
FRANCE
MONACO
ANDORRA (FR-SP)
SPAIN
AZORES (PORT)
PORTUGAL
GIBRALTAR (UK)
CEUTA (SP)
MELILLA (SP)
MADIERA (PORT)
CANARY ISLANDS (SP)
WESTERN SAHARA
(MOR)
MOROCCO
CAPE VERDE
ISLANDS
MALI
THE GAMBIA
GUINEA-BISSAU
BURKINA
SIERRA LEONE
LIBERIA
TOGO
EQUATORIAL GUINEA
SAO TOME & PRINCIPE

ST. HELENA AND
DEPENDENCIES (UK)

UNITED STATES

CANADA

ST. PIERRE-MQ. (FR)

PACIFIC OCEAN

UNITED STATES

BERMUDA (UK)

ATLANTIC OCEAN

UNITED STATES

MEXICO

BAHAMAS
CUBA
CAYMAN
ISLANDS
(UK)
BELIZE
GUATEMALA
EL SALVADOR
HONDURAS
NICARAGUA
COSTA RICA
PANAMA

PUERTO RICO
VIRGIN ISLANDS (US)
BRITISH VIRGIN ISLANDS (UK)
ANGUILLA (UK)
ST. KITTS-NEVIS
ANTIGUA & BARBUDA
MONTSERRAT (UK)
GUADELOUPE (FR)
DOMINICA
MARTINIQUE (FR)
ST. LUCIA
ST. VINCENT & THE GRENADINES
BARBADOS
TRINIDAD & TOBAGO
TURKS & CAICOS
(UK)
JAMAICA
HAITI
DOM. REP.
NE. ANTILLES (NE)
ARUBA (NE)
GRENADA

VENEZUELA
GUYANA
SURINAME
FRENCH GUIANA
(FR)
COLOMBIA
ECUADOR
PERU
BRAZIL
BOLIVIA
PARAGUAY
CHILE
ARGENTINA
URUGUAY

AMERICAN SAMOA (US)
NIUE (NZ)
COOK ISLANDS (NZ)
FRENCH POLYNESIA (FR)
RAPANUI/EASTER ISLAND
(CHILE)
PITCAIRN ISLANDS (UK)

FALKLAND ISLANDS (UK)

FREEDOM
HOUSE

FREE ☐   PARTLY FREE ▨   NOT FREE ▮

JANUARY 1997 ©FREEDOM HOUSE

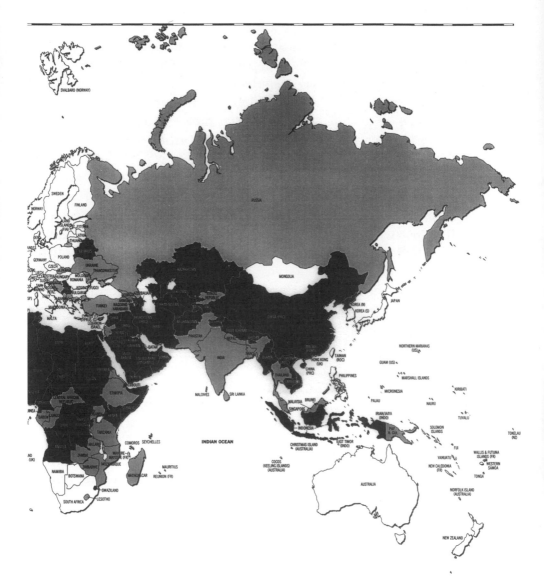

The Map of Freedom

system—a move much opposed initially by Chilean unions—
which now encompasses all workers. Pinochet also instituted
rules recognizing the right of private citizens to hold property,
along with a supportive legal system. The Chilean transformation from a society with virtually no political freedom to
one now regarded as wholly free shows that authoritarian
leadership can occasionally hasten a nation along the freedom path.

New Zealand also ranks as a nation that has long enjoyed political freedom. But its degree of freedom has risen in
recent years for reasons that reflect far less draconian events
than occurred in Chile. New Zealand was not under the heel
of any dictator. Rather, it had been run by left-leaning politicians who promoted pervasive government interference in
economic and social matters and dismissed the potential benefits of laissez-faire. But things changed in the mid-1980s. An
editorial writer for the *Wall Street Journal* later described the
transformation this way: "This little Prometheus unchained
itself from a rock of high taxes, high tariffs, heavy welfare
burdens and pro-union labor laws."[3] And now, as a result,
New Zealand boasts one of the highest ratings assigned to
any nation by Freedom House, far higher than its rating as
recently as 1985.

This welcome change, I should point out, reflects not only
New Zealand "unchaining" itself from various regulatory and
other burdens, but from a gradual, deliberate strengthening
of institutions crucial to sustaining a free, prosperous society.
One such step was the passage in 1989 of legislation increasing the authority of the nation's central bank. The Reserve
Bank of New Zealand Act mandates that the nation's central
bank maintain price stability—period. Under law, there now
must be no interference with monetary policy by political

3. "Kiwi School of Economics," *Wall Street Journal*, December 14, 1994.

leaders, whose inclination in the past was all too often to inflate the currency.

With this legislation, New Zealand's rate of inflation has held at enviously low levels in the 1990s, in happy contrast to annual rates ranging above 15 percent before passage of the 1989 act. Among freedom's most insidious enemies, I need hardly add, is rampant inflation. Annual inflation of 15 percent will halve a currency's buying power in less than five years!

Yet another encouraging story has recently been unfolding in the Philippines, where the nation's congress has declared Subic Bay, once a huge U.S. Navy base, a duty-free trade zone. More than 45,000 Filipinos work there in nonmilitary jobs, ranging from servicing cruise ships to assembling telephones. That's far more than the 13,000 jobs that existed at the Subic Bay base in 1992, just before the U.S. Navy departed. It even exceeds the 1992 total if one includes some 25,000 bar girls who lost their jobs. This growing prosperity at Subic Bay is matched at other localities across the nation. By no coincidence, the Philippine economy has recently been expanding at more than 7 percent annually. That compares with no growth at all a few years ago.

This striking improvement is widely attributed to the policies of Philippine President Fidel Ramos. A former soldier, he has managed to make peace with Muslim secessionists and pacify a military establishment known to stage periodic coups. When President Ramos took charge, much of the advice that he received was to exert tougher discipline and, in the process, roll back democracy. Instead, he has encouraged broader political freedom and loosened the powerful grip of a few Filipino families, including that of his predecessor Corazon Aquino, that had long held monopolistic control over much of the business activity transacted in the Philippines.

Roberto De Ocampo, the Philippine minister of finance, recently described the Ramos strategy this way: "We are dedicated to formulate a new model of Third World economic development that combines growth with equity, modernization with environmental consciousness—all in the context of democracy."[4]

The success of the Ramos regime is reflected in the nation's work force. A Hong Kong research firm, Political & Economic Risk Consultancy Ltd., recently conducted a survey of international executives to assess the quality of workers in dozens of nations around the world and those in the Philippines were ranked near the top, in fourth place.

Increasingly, Majorities Rule

Consider these far-flung scenes, all captured by roving photographers and subsequently printed in the *New York Times* on August 4, 1996:

- In Mongolia, a man with a weather-beaten face stands by as his son and a fellow nomad enter a tent on June 30 to cast their votes in a remote area some 150 miles east of Ulan Bator.
- In Chad, a voter garbed in flowing white robes picks up his ballot, kisses it, and prays before depositing it on June 2 in that African nation's first free election.
- In the Middle East, a Jewish settler on the West Bank, a submachine gun slung across his shoulder, casts his ballot in Hebron on May 29 during Israel's national election.
- In Taiwan, retired army veterans, all elderly and in wheelchairs, patiently line up near a voting station in Tapei to cast their ballots on March 23 in the country's first direct presidential election.

4. Peter Waldman, "Success Story," *Wall Street Journal*, October 4, 1996.

As such vignettes indicate and the headline on the August 4, 1996, *Times* article declares, "Globally, Majority Rules."[5]

This proliferation of democracies around the world is particularly heartening when one reflects that the past century provides no clear-cut instance of one democracy waging war against another democracy. As a recent survey concludes, "Democracies are better trading partners and rarely fight one another."[6]

You may recall, in this regard, the collective sigh of relief breathed in the United States and elsewhere on the morning of July 4, 1996. This, of course, was when it became clear that Russian President Boris Yeltsin had defeated, by an unexpectedly wide margin, his main challenger, the Communist leader Gennadi Zyuganov.

The trend toward democracy extends even to Arab countries, which refutes the notion that Islam and democracy cannot mix. Recently, for example, Algeria for the first time held a multi-party presidential election, with four candidates vying for the support of the nation's 16 million voters. Previously, the nation had been ruled by a single political party since its independence from France in 1962.

To be sure, not all the democracies that now engird the globe are completely free. For citizens to enjoy complete political freedom, in addition to the secret ballot box and majority rule, there must exist institutions designed to preserve orderly, effective democratic rule. These include such pillars as an impartial judiciary whose decisions are respected and enforced, an effective system of tax collection, and an incorruptible military presence. Also crucial, of course, is a free press,

5. Barbara Crossette, "Globally, Majority Rules," *New York Times*, August 4, 1996.
6. Joseph Nye and William Owens, "America's Information Edge," *Foreign Affairs*, March–April 1996.

along with related independent organizations whose research helps inform the electorate and points political leaders toward wise policies.

It is heartening to observe that even in a democracy as new and fragile as Russia's, steps are under way to bolster the democratic process. While not yet on a par with the press in western Europe, Russia's press is vastly freer than in the Soviet years. And *Pravda*, for decades the mouthpiece of communism, is finally out of business.

At the same time, Russia is taking steps to ensure that taxes are collected—an unpleasant but essential task that democratic governments must undertake to function properly. The government is clamping down on tax evasion, which has been rampant. Estimates show that only some 60 percent of taxes owed in Russia have actually been collected in recent years. Among other recent measures, the government has established a commission to work out schedules by which companies may pay overdue taxes. Enterprises that resist are being hauled into court, where stiff fines and other penalties are applied.[7]

Russia's new drive to enforce its tax system has even led to the 1996 opening in Moscow of a new store called Arrears. It sells at bargain prices wares seized by Russian tax collectors from businesses in arrears on taxes. Each morning Moscovites line up outside the store for an early chance to buy such items as down coats at the equivalent of $17, red corduroy pants at only $2, denim backpacks at $3, and even a used bus at $19,000.[8]

In Africa, meanwhile, the small nation of Malawi now boasts more than thirty newspapers, up from only one as

7. Neela Banerjee, "Russia to Get Tough With Tax Evaders," *Wall Street Journal*, July 31, 1996.

8. Alessandra Stanley, "The Tax Man's Booty Stocks a Bargain Store," *New York Times*, September 14, 1996.

recently as 1993. And in Mexico, governmental press controls are loosening under the leadership of President Ernesto Zedillo. *Reforma*, a newspaper based in Mexico City, now prohibits its reporters from indulging in what was once a common practice—accepting payments from government officials in exchange for friendly coverage. *Reforma* has also initiated quarterly reports by a polling unit that gauges how the public rates Zedillo. A poll in mid-1996, for example, found that 57 percent of Mexicans disapproved of the president's performance, the sort of news that past Mexican governments would not have tolerated. In 1972, for example, government agents destroyed the entire press run of a newspaper deemed unfriendly to the incumbents.[9]

Around the world, fewer journalists are being abused. In 1995, according to Freedom House, 310 journalists were detained by governmental authorities, 32 were kidnapped or disappeared, and 174 suffered physical harm. These numbers may seem high to people residing in the United States or Western Europe, but they mark a decided improvement from a year earlier, when 345 journalists were detained, 38 were kidnapped or disappeared, and 275 were roughed up.

Particularly heartening has been the strengthening of the press in South Africa. During that nation's recent, historic move away from apartheid, a special commission was set up to safeguard the integrity of the South Africa press corps. And now, after Nelson Mandela's decisive victory, an Independent Broadcasting Authority limits the government's control over broadcast media and the issuance of broadcast licenses.[10]

A further support for South Africa's new democracy is the nation's Constitutional Court. This institution, akin to the

9. Sam Dillon, "Mexico's New Press Boldness," *New York Times*, August 15, 1996.

10. Thomas R. Lanser, "The Press in South Africa," *Freedom Review*, March–April 1996.

U.S. Supreme Court, recently overruled a Mandela decree that would have apportioned voting districts in a manner deemed partial to the incumbent party.

Spreading Individual Rights

Topping the list of yardsticks by which Freedom House measures the level of freedom in a particular nation is the matter of freedom to hold property. To gauge this freedom, the questions raised by Freedom House may seem unnecessary to someone residing in, for example, the United States. But in much of the world they remain matters of great concern. Among them: Is the right to own property recognized by law? Is the right to intellectual property protected? Does the legal system support the right to property? Are there restrictions on selling, exchanging, or dividing property?

While no single statistic can confirm the spread of the freedom to hold property, anecdotal evidence gathered around the world strongly points to such a trend.

• Colombia's 1991 constitution explicitly protects individual property rights against governmental expropriation. Around the same time, the nation's port system and banking and telecommunications industries were privatized.

• In nearby Peru, legislation enacted in 1993 recognizes the right to private ownership of property and, reversing an earlier pattern, there has been no expropriation of property since that time. Meanwhile, key Peruvian industries are in the process of being privatized, including oil, steel, electricity, railroads, and mining. By no coincidence, the government-owned share of the economy has shriveled to less than 5 percent at present from 35 percent in 1992.

• In the Czech Republic, shortly after the fall of communism there, the new government transferred nationalized real estate to the control of municipalities where the property was located. Residents could then apply to buy their homes from the municipalities at discounted prices. In this way, most housing is now in private hands and there no longer are any restrictions on buying or selling real estate. The government has also recently agreed to safeguard intellectual property rights.

• In Latvia, another former communist nation, control of housing and land has been restored to private owners. In addition, legally transferable rights to private property have been established, along with the right of individuals to structure their property holdings as they choose. Freedom House now assigns Latvia the same high freedom rating that it gives to such nations in the West as the United States and Germany.

• In Slovakia, also once a part of the Soviet empire, the right to own tangible as well as intellectual property is now recognized by law. One result is that some 400,000 acres have been returned since 1993 to some 24,000 private owners. Moreover, restrictions on private land use have been lifted.

The rights of individuals are strengthening even in nations where political freedom remains generally weak. In Vietnam, for example, private property ownership is still banned, but under a 1993 land law, individuals now have the right to transfer, lease, mortgage, and inherit land-use rights.

Uzbekistan, with a population of nearly 23 million, has done little to dismantle Soviet-era controls since its independence from the former Soviet Union and its general freedom rating is near the bottom on the Freedom House list. Yet, the government recently issued a decree legalizing the private ownership of land used by small businesses. In addition, the nation's farms are in the early stages of privatization. And a presidential order in March 1994 opened the way for the construction of private housing.

Such steps may not seem impressive to people fortunate enough to live in nations where extensive freedom has long been the rule. But they do serve as a clear indication that political freedom is continuing to expand around the world—

even in places where only a few years ago concepts like private housing or inherited land rights were unthinkable.

Of course, a longer look back in history underscores the truly astounding strides that have been made in the progress of personal freedom. Less than 150 years ago, slaves were still being bought and sold in the United States. Five years after slaves were freed in 1865, at the end of the Civil War, an amendment to the U.S. Constitution guaranteed their right to vote. (Interestingly, it took another 50 years for an amendment allowing American women of any color to vote.)

Expanding the rolls of eligible voters, to be sure, constitutes an important element in increasing political freedom in any nation. At one time, it was considered entirely reasonable that only U.S. land owners, men with a stake in the country, should be allowed to vote. Even in nations where the democratic process is well-established, anomalies have existed well into the recent past. In Switzerland, a democracy since its founding in 1291, women were disenfranchised until the early 1970s, when most Cantons agreed to allow them to vote. In the United States, it wasn't until 1971 that a Constitutional amendment lowered the voting age from 21 to 18, recognizing the justice in "if you're old enough to fight, you're old enough to vote."

Hotly contested elections among opposition political parties have been the case in South Africa for many years, but under the apartheid system only the white minority was allowed to cast ballots. Finally, on May 10, 1994, Nelson Mandela, a black leader and long-time political prisoner, was inaugurated as president, following the first free popular election in South Africa's three-hundred-year history.

The spread of political freedom around the world is truly a great blessing, for as we shall see, it begets and augments other freedoms. Indeed, it constitutes a crucial part of a broader, irreversible process that has led—and continues to lead—to better, more fulfilling lives for everyone.

7.

Economic Freedom

"We have a capital market that is a great national asset, like the wheat fields of Kansas and the Grand Canyon."
—Adam Smith, *Supermoney*

"[Our] survey found that in most nations privatization of state enterprises and the removal of other direct restraints on economic life have become the accepted wisdom."
—Richard E. Messick, director of Freedom House's 1995 World Survey of Economic Freedom, *Wall Street Journal*, May 6, 1996

- There are some 10 million private businesses in the United States today, four times the number as recently as 1970.
- More than 10,000 U.S. institutions, with more than $5 trillion under management, use the New York Stock Exchange to buy and sell securities.
- Three billion people worldwide live in free-market economies.

People's Capitalism

The new freedom that now reaches so much of the world has economic and financial, as well as political, dimensions. From the Czech Republic to Chile to New Zealand to Malaysia, nations are pursuing, with mounting success, policies that encourage the forces of the marketplace and individual enterprise. They are stripping away stultifying economic restraints, from governmental regulations to burdensome taxes to rules that hamper international trade and investment. And they are instituting safeguards for private investors, such as tighter requirements for corporate financial reporting.

Only a decade ago, barely 1 billion of the world's population were guided by economic rules encouraging market forces, while today some 3 billion people are ensconced within the free market's spreading tent. As one analysis concludes, "The pace at which capitalism has rolled through developing economies is breathtaking."[1]

All of this is in marked contrast to what prevailed only a couple of decades ago, when much of the Third World operated under various constraining forms of socialism and central economic planning. Vaclav Klaus, the Czech prime minister, spoke not just for his own country but for many nations when he declared in 1991 that "we need an unconstrained, unrestricted, full-fledged, unspoiled market economy, and we need it now. And," he went on, "we want to achieve the transition from a state-dominated economy to an economy based on the private sector, private initiative and private enterprise."[2]

There is good reason for such sentiments. A recent survey by Freedom House rated eighty-two nations according to the freedom of their economies. Forty-nine of the group were considered "free" or "partly free," while the remaining thirty-three were deemed "mostly not free" or "not free." The survey went on to note that the forty-nine, though representing only 24 percent of the world's population, accounted for 86 percent of the world's gross domestic product. The other thirty-three, with fully 66 percent of the world's population, accounted for only 13 percent of global GDP.

Further evidence of increasing economic freedom shows up in a report by the Fraser Institute that weighs conditions in more than one hundred nations. Studying such questions as the top levels of marginal tax rates and the extent of government controls over foreign investment and currency transactions, the report finds, by and large, that "a significant

1. "Financing World Growth," *Business Week*, October 3, 1994.
2. Roger Douglas, *Unfinished Business,* (Auckland, New Zealand: Random House, Ltd., 1993).

amount of liberalization has occurred since 1985." In 1985, for example, as many as forty-eight of the countries surveyed imposed top marginal tax rates of 60 percent or more, while a decade later only ten imposed such high rates. In 1985, moreover, thirty-eight countries imposed such restrictive exchange controls that the premiums on accompanying black markets were at least 25 percent. By 1995, however, only eleven nations still imposed controls severe enough to trigger such high black-market premiums.[3] Other evidence of a global shift toward much greater economic freedom ranges from increasing privatization of government enterprises to the lifting of wage and price controls.

Even in Russia, once a paragon of top-down statism, what exists now can be described, at the least, as frontier capitalism. It is hardly to be envied by those of us fortunate enough to reside in the United States or Western Europe for example, but it may constitute a necessary stage in Russia's painful progress toward a prosperity founded on the market forces of supply and demand.

The demise of communism in Russia and elsewhere in what was once known as the Soviet bloc was inevitable. Central planning leaves little or no room for personal incentive and initiative. Rather, it assumes—wrongly—that people will work selflessly to construct a better, fairer society. But human nature, by and large, doesn't work that way. Even the most energetic and able central planners—admittedly, a rare breed—are incapable of making the countless decisions needed to manage a diverse, sophisticated economy. Indeed, the more diverse and sophisticated a nation's economy becomes, the more impossible central planning becomes.

The fallacy in central planning, including the toned-down version called industrial policy, was well described years ago by the Austrian economist Friedrich Hayek. Countering

---

3. "Economic Freedom of the World, 1975–1995," *The Fraser Institute*, 1996.

> Sidewalk vendors are everywhere. . . . Middle-aged men and women sell books, Manhattan-style, on sidewalk tables laden with . . . dictionaries, cookbooks, Sophia Loren's beauty advice, translated into Russian [while] younger vendors man refrigerated pushcarts, peddling ice cream and cold drinks. [6]

Greater economic freedom is also benefitting many nations that, unlike Russia, have eschewed the path of communism. An example is New Zealand, where in 1984 a new government abandoned its predecessor's interventionist economic policies, scaled back regulatory restraints on business, reined in "entitlement" spending, cut taxes, and, to curb inflation, bolstered the central bank's independency. Once a basket case, New Zealand's economy is now among the most vibrant in the world.

Today's spreading appreciation of the marketplace extends even to African nations still struggling to keep living standards from sinking after decades of rule by autocratic leaders preaching the virtues of state-run socialism. "Today you won't find a single African head of state who stands on a podium and declares: 'I am a Marxist,'" says Tei Mante, a Ghanaian who heads the African office of the International Finance Corporation, an affiliate of the World Bank. Instead, he says, "all the talk is about floating currency, private enterprise and getting hold of capital."[7]

A notable illustration of this transition away from central control and toward free markets exists in Chile, not so long ago an economic cripple but now boasting one of the world's healthiest economies, with brisk growth and sharply reduced inflation. Reflecting Chile's new circumstances is Vincent A. Russo, the general manager of Santa Fe Forestal y Industria, a $460 million pulp plant in Nacimento. Russo

---

6. Ibid.
7. John Darnton, "In Poor, Decolonized Africa Bankers Are New Overlords," *New York Times*, June 20, 1994.

proudly declares that now "we have a country committed to free enterprise, where there are established rules, a central bank you can deal with and zero bribery."[8]

Such change is already affecting investor decisions mightily, as evidenced by the flow of global capital. In 1990, some $31 billion of foreign direct investment went to the developing world, while $176 billion went to developed economies, but within three years the flow to developing nations had reached some $80 billion and the rise is continuing.

By no coincidence, there is widespread privatization throughout the developing world of enterprises that not long ago were government-run. United Nations data show that as recently as the mid-1970s, the number of nationalizations of private businesses around the globe outstripped the number of privatizations by a ratio of roughly five to one. But lately the pattern has reversed, with a ratio of approximately one to five.

Argentina provides a striking illustration of this turnabout. In recent years, as many as fifty-one state-owned enterprises have been sold for a total of $5.6 billion in cash. The privatized sectors, no longer draining the governmental coffers, include oil, electricity, gas, shipping, and railroads. In the process, the nation's foreign debt has shrunk by nearly $12 billion, while some $1.3 billion in financial liabilities has been assumed by new private owners.[9]

The trend toward privatization was evident at a meeting in Madrid in the fall of 1994 of the World Economic Development Congress, a private, multinational organization. Addressing the conference, Kenneth D. Brody, head of the U.S. Export-Import Bank, reported a "cataclysmic change" in the sort of export financing being undertaken by his organization. "Two years ago, most of our business was with sovereign

8. Nathaniel C. Nash, "A New Discipline in Economics Brings Change to Latin America," *New York Times,* November 13, 1991.

9. William Ratliff and Roger Fountaine, "Argentina's Capitalist Revolution Revisited," *Hoover Institution,* 1993.

governments," he recalled, while "this year, over half will be with private sectors in developing countries, and the percentage is growing."[10]

A particularly memorable example of how times are changing occurred in Jamaica in 1995, when the government there put up for sale to private investors the nation's major airport. Located in the capital city of Kingston, the facility is named after the late Norman Manley, a leading Jamaican socialist and the father of Michael Manley. The younger Manley, a former Jamaican prime minister, once espoused the virtues of socialism himself, as well as a high regard for Fidel Castro's communist regime in nearby Cuba. But he no longer holds such views. Instead, he now backs economic policies that encourage a relatively free, competitive marketplace.

Second to none in the effort of developing nations to scale back government and encourage private investment is Bolivia, whose president, Gonzalo Sanchez de Lozada, recently launched a program to privatize a broad range of government-owned facilities. Bolivia, however, has added a special twist to its privatization program. Payments from private investors will go directly into modernizing facilities being privatized. This, the president explains, helps keep the investment money from simply going into the government budget for use "by politicians for their own pet projects."[11] In addition, the plan is set up so that 50 percent of the assets are owned by the Bolivian people, through privately-run pension funds.

Privatization programs are under way throughout South America. One privatization target is transportation, which isn't surprising considering the vastness of South America and the desire of many nations there to integrate their

10. George Melloan, "Developing Nations Are Getting Off the Dole," *Wall Street Journal*, October 3, 1994.
11. Gonzalo Sanchez de Lozada, "Bolivians Challenge the Hemisphere to Dream Big," *Wall Street Journal*, November 11, 1994.

economies more closely. Brazil, for instance, has turned to private companies to double the width of the country's much-used main highway, a 265-mile link between Sao Paulo and Rio de Janeiro. Under the terms of the $1 billion project, the contractors will also have the right to levy tolls.

In Chile, similarly, the Public Works Ministry recently sought bids from private firms on a plan to enlarge one thousand miles of the Pan American Highway into a divided, four-lane road with toll booths positioned every one hundred miles or so. And since 1991, Argentina has given private construction companies various concessions to rebuild some six thousand miles of busy highways.

Elsewhere in South America, Peru in late 1994 held its first auction of shares in privatized government enterprises. Though the offering was limited to citizens with relatively low incomes, some 3,000 people snapped up the available shares in less than three hours, and 6,600 low-income investors gobbled up a subsequent offering in only forty-five minutes. By mid-1996, this exercise in people's capitalism had grown to involve more than 100,000 Peruvians, and the government is aiming at a half million such investors by the year 2000.[12]

Privatization continues apace in parts of the world where transportation may pose less of a challenge than in South America. The Czech Republic has completed a privatization program in which state-owned businesses were sold to some 6 million individuals through an exchange of coupons for ownership shares. Some 80 percent of the government's assets—in all, 1,849 enterprises—wound up in private hands. The program is regarded as a model for other public-to-private conversions in Eastern Europe's bloc of formerly communist nations.

---

12. Carol Graham, "People's Capitalism Makes Headway in Peru," *Wall Street Journal,* April 19, 1996.

The Amazing Growth of Global Markets

With greater economic freedom has come an exponential increase in the international flow of capital. The supply is not
unlimited, of course, but with modern communications and
the split-second electronic transfer of funds, billions now can
move instantly from computer screen to computer screen and
from continent to continent. A few statistics underscore this
rise in capital mobility. As recently as 1989, some $700 billion
of financial settlements were cleared each day through the so-
called CHIPS (or Clearing House International Payments
System) wire at the New York Clearing House. Now, the daily
total exceeds $1.1 *trillion*, an increase of more than 50 per-
cent. David Hale, chief economist of Kemper Corp., sees in all
of this the creation of a "new world economy" in which, for the
first time in history, "virtually every country on the planet
has a market-oriented economic system, and is attempting to
be a player in the global marketplace for goods and capital."[13]

In his aptly titled 1992 book, *The Twilight of Sovereignty*,
Walter B. Wriston, the former Citicorp chairman, observes
that the world's main economies are "now tied together in a
single electronic market moving at the speed of light." Wriston adds that "no matter what political leaders do or say, the
screens will continue to light up, traders will trade, and currency values will continue to be set not by sovereign governments but by global plebiscite."[14]

The spread of economic freedom has brought a proliferation of markets on which investors, small as well as large,
may buy shares of stock and other financial instruments.
Such markets are of long standing in nations like the United
States, Britain, France, Germany, Switzerland, and the

13. David Hale, "Rethinking the World," *Barron's,* August 22, 1994.
14. Walter B. Wriston, *The Twilight of Sovereignty* (New York: Charles Scribner's Sons, 1991).

Netherlands. They also exist in such places as Japan, Hong Kong, Singapore, Thailand, Indonesia, Malaysia, the Philippines, and Taiwan, as well as such diverse countries as Canada, New Zealand, Spain, South Africa, and Italy. And still other securities markets have emerged in such newcomers to people's capitalism as Russia, the Czech Republic, and mainland China, places where socialism and communism are dead or dying.

Consider China. In December 1990, the Shanghai Securities Exchange opened in that Chinese city, becoming modern China's first recognized stock exchange. Only a few months later, in April 1991, the Shenzhen Stock Exchange opened. A year later, the Chinese government established the China Securities Regulatory Commission to supervise the two exchanges, whose lists of shares have grown as more and more of China's government-owned enterprises are privatized and seek to raise money through issuing stock to investors.

And what a marketplace China has become! With its economy expanding at rates of close to 10 percent a year, the rapid industrialization of this Asian giant will require awesome amounts of private investment capital, as well as technical and managerial help from abroad. According to the Asian Development Bank, infrastructure needs in all of developing Asia—such as for telecommunications, electric power, transportation, water supply, and sanitation—will approximate $1 trillion over the remaining years of this century, and China will account for more than half of this vast demand. Just to turn the lights on in the thousands of new Chinese factories springing up *each year* requires adding 12,000 megawatts of power-plant capacity. For perspective, 18,000 megawatts of generating capacity powers the whole of Southern California.[15]

15. Patrick E. Tyler, "Awe-Struck U.S. Executives Survey the China Market," *New York Times*, September 2, 1994.

Helping to meet China's huge demand for capital, Japan's direct investment there now exceeds $2 billion yearly, double the rate as recently as 1992 and four times the 1991 amount. And China's hunger for capital is mirrored in dozens of other emerging areas—elsewhere in Asia, in Latin America, and in parts of the former Soviet bloc.[16]

In all these nations, for all their recent economic gains, a vast potential of unfilled needs remains. In Mexico, only 31 percent of households yet have clothes dryers and only 19 percent have microwave ovens. In the United States, by comparison, 74 percent of households have clothes dryers and 44 percent have microwaves. Mexico's economy and its financial markets, I should add, are among the more highly developed to be found in the Third World.

Demographics point to burgeoning demand in emerging markets. Some 85 percent of the world's population of 5.5 billion resides in these nations and the average age structure of this group is appreciably younger than for the developed nations. In India, some 64 percent of the population is under the age of thirty and in Indonesia the percentage is 65 percent. By comparison, the rate in the United States is only 46 percent and just 40 percent in Japan.

The rapid economic growth of many emerging nations is also benefitting economies in the developed world. "Latin America has become the fastest growing market for U.S. exporters," says Arturo Vera, an economist for the Inter-American Development Bank in Washington. As a result of "macroeconomic reforms" within the region and liberalized trade policies, he says, U.S. merchandise exports to Latin America crossed above the $80 billion mark for the first time in 1993, roughly double the comparable level in 1980. U.S. exports to Brazil alone surged 35 percent in 1994.

16. Bank for International Settlements, 1994 Annual Report.

Looking ahead, the Office of the U.S. Trade Representative projects that U.S. exports to Latin America will reach $232 billion (in terms of 1994 dollars) by the year 2010. That would be up from an actual level of $88 billion in 1994. The trade office estimates a similarly sharp gain in U.S. exports to the emerging nations of East Asia. The volume is expected to reach $248 billion by 2010, up from $94 billion in 1994.

Today, U.S. exports to Canada and Japan combined are roughly twice as large as its exports to Latin America. But by 2010, U.S. sales to Latin America are expected roughly to match its exports to Canada and Japan. Similar strong gains are anticipated in U.S. exports to other Third World areas, as well as to countries that once were communist.

Such forecasts assume—correctly, in my opinion—that the global shift toward greater economic freedom will continue and that reforms recently implemented in former communist nations will persist, though setbacks, such as the recent peso crisis in Mexico, will doubtless occur from time to time.

In many emerging nations, efforts are being stepped up to reduce impediments to private investments. In October 1991, Argentina ended a 36 percent capital gains tax and removed a three-year holding period that had been required before foreign investors could repatriate capital. About the same time, Brazil eliminated a similar holding requirement and slashed tax rates on dividends and capital gains. Elsewhere, South Korea in 1992 opened its market to direct stock purchases by foreigners, though foreign ownership of listed companies is limited to 10 percent. This move was on the heels of Thailand's decision to remove a 25 percent capital gains tax and ease its rules for the repatriation of capital. Around the same time, India lowered its tax rates on capital gains and dividends. And in mid-1994, Indonesia, once leery of foreign investment, scrapped a rule that foreigners investing there must have an Indonesian partner. Now, 100 percent foreign ownership is permitted.

Impediments to trade also are being hauled down in many emerging markets. Once highly protectionist, the Philippines plans to reduce all import tariffs to a maximum rate of 5 percent by 2003. Leaders of the eighteen-nation Pacific Rim bloc intend to turn the region into a huge free trade zone, the world's largest, by 2020.

In South America, Brazil, Argentina, Uruguay, and Paraguay have joined together in a free-trade bloc called Mercosur, encompassing 200 million people. The volume of intra-bloc exports reached a record $12 billion in 1994, four times the comparable amount only four years earlier.

The benefits of such measures are already becoming evident. As Michel Camdessus, the International Monetary Fund's managing director, has put it with regard to Latin America, "The benefits are ample and tangible and offer the prospect of long-term development with a higher standard of living for people today and for generations to come."[17]

Abounding Opportunities to Invest

The world's great financial and foreign-exchange markets have long served relatively wealthy investors wishing to buy and sell all sorts of securities and currencies in their efforts to preserve and increase capital. But now, through an expanding variety of financial intermediaries, these same markets increasingly are serving investors of quite limited means. These markets also serve, of course, to help corporations raise capital through initial public offerings of stock, or IPOs. In addition, thousands of other, smaller companies raise capital directly from so-called venture capitalists willing to provide funds in exchange for early stakes in new concerns.

17. Michel Camdessus, "Social Dimensions of Economic Restructuring," *IMF Survey*, December 14, 1992.

In all, there are some 10 million private businesses in the United States today. This is an awesome total, more than four times the number as recently as 1970. New businesses are springing up, moreover, at the remarkable rate of about 700,000 yearly, up from 264,000 in 1970. Some 400,000 of these businesses, I should add, attract about $50 billion annually from outside investors. Clearly, there is no shortage of businesses in which to invest for the more than 51 million Americans who now own stock in individual companies or shares in stock mutual funds, whose diversified holdings allow all investors a degree of diversification once available only to the well-off.

The table below traces the steep rise of America's shareholder population in less than a quarter of a century. Note, particularly, the decline in the median age of these shareholders and their rising incidence in the nation's adult population.

There are many routes by which these investors may partake in the ownership of capital, but the best known is the New York Stock Exchange. More than 10,000 U.S. institutions, with more than $5 trillion under management, use the Big Board, as the NYSE is commonly called, to buy and sell securities. The more than 2,600 companies whose shares are traded on the Big Board range from relatively small enterprises to giants whose yearly sales exceed the total economic output of some not-so-small nations. The list also includes

People's Capitalism

	1975	1980	1985	1990
Shareholders (thousands)	25,270	30,200	47,040	51,440
Shareholder incidence in adult population	1 in 6	1 in 5	1 in 4	1 in 4
Median age	53	46	44	43

SOURCE: New York Stock Exchange

more than 200 firms whose headquarters are outside the United States.

On an average day in 1994, as many as 291.4 million shares of stock were traded on the Big Board, a 10 percent increase from a year earlier and a record. Total share volume of some 74 billion shares was also at a record level, as was the exchange's full-year trading volume, valued at nearly $2.5 trillion.

The next table tracks the growth in trading volume of shares listed on the Big Board and six other U.S. stock markets: the Pacific Stock Exchange (PSE), the Chicago Stock Exchange (CHX), the Philadelphia Stock Exchange (PHLX), the Boston Stock Exchange (BSE), The Cincinnati Stock Exchange (CSE), and the National Association of Securities Dealers (NASD), also known as Nasdaq.

There are still other stock exchanges in the United States, including the American Stock Exchange, which is smaller than the Big Board but also based in New York, and the so-called Nasdaq—short for the National Association of Securities Dealers Automated Quotation System. On the Nasdaq, investors may freely trade shares of companies that

Growth in Volume of Shares Traded 1984–1996

Consolidated volume by market (millions of shares)

	NYSE	PSE	CHX	PHLX	BSE	CSE	NASD
1996	87,485	2,341	3,655	1,389	1,400	1,826	8,458
1994	73,578	1,901	3,226	1,216	1,108	1,369	6,473
1992	51,529	1,882	2,806	1,042	939	647	4,145
1990	39,926	1,465	2,353	915	815	313	2,323
1988	40,850	1,340	2,633	632	593	252	1,034
1986	35,680	1,541	2,657	701	599	166	1,084
1984	23,071	886	1,865	468	265	52	799

SOURCE: New York Stock Exchange

typically are younger and smaller than firms whose shares are traded on the Big Board. Unlike most stock exchanges, the Nasdaq has no trading floor. Rather, it is simply a grouping, via telephones and computers, of some five hundred securities firms that trade shares with one another for their clients.

Meanwhile, the ranks of stockbrokers continue to expand. In the United States alone, employment in the securities industry, at 367,500 as recently as 1985, now exceeds 510,000, an increase of nearly 40 percent in a single decade.

To see how investors of limited means benefit from changes in the investment scene, let's imagine that John Smith, a neophyte investor, wishes to buy one hundred shares of the stock of XYZ Corp., a well-known and giant enterprise. John's broker tells him that this purchase will cost $15 a share, including the broker's commission. With his resources limited, John wisely decides that to do this would place too much of his capital within a single basket—namely, XYZ stock. So he opts instead, with his broker's help, to diversify his investment through a mutual fund.

Typically, mutual funds use money supplied by many investors to buy a wide assortment of stocks, as well as perhaps bonds, currencies, or financial instruments. By placing his $1,500 in a large mutual fund, John protects himself against any unexpected trouble in a single stock. If XYZ shares were to plunge, for example, this wouldn't greatly hurt his overall investment in a mutual fund, even if its holdings included XYZ. This is because the fund's holdings would likely be spread among scores of stocks, including many lesser-known companies that John may never have heard of, as well as a number of foreign-based concerns. The fund's portfolio may also encompass various corporate bonds and short-term debt instruments, which are tantamount to cash.

As a mutual-fund participant, John will wind up owning, in effect, small pieces of whatever institutions the fund as a whole may choose to hold. Also, John will have the additional benefit of a professional investment management team at the mutual fund, rather than having to rely simply on his broker.

The chart below shows that many investors, like John, are opting to diversify through mutual funds. It traces the sharp rise in recent years in the share of U.S. consumers' financial holdings placed in mutual funds.

The chart on the right shows the enormous rise in recent decades in the asset value of mutual fund holdings, as well as in the number of mutual funds in the United States.

The freedom of choice that investors such as John Smith now enjoy extends well beyond the stock market, whose vagaries can send large ripples through even the largest and most diversified mutual funds.

John can also invest through other sorts of financial intermediaries. Insurance companies offer an illustration. For a fee or regular premium, insurance companies will pay specified sums to people who take out policies with them. These amounts typically are in the form of insurance against calamities, but they may also be used to provide a steady

Mutual Funds as a Percentage of Individual Financial Holdings

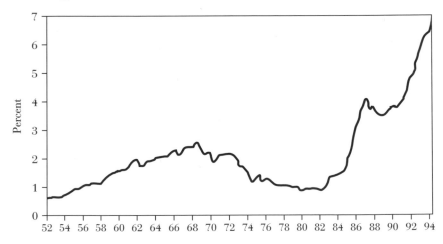

SOURCE: Federal Reserve Board

Growth in U.S. Mutual Funds, 1980–1996

	Assets of Mutual Funds (billion $)	Number of Mutual Funds
1996	3,540.3	6,270
1995	2,820.3	5,761
1994	2,161.5	5,357
1990	1,066.8	3,105
1985	495.5	1,528
1980	137.2	564

SOURCE: Investment Company Institute

income flow, called annuity payments, when policy holders retire or reach a certain age.

Like other financial intermediaries, insurance companies invest money that they receive from policyholders in the capital markets, primarily buying instruments such as bonds and mortgages that produce a relatively large, steady cash return. If John Smith feels that he won't need any of that $1,500 for decades and wants to ensure that he'll get regular payments when he retires in forty years, he may decide to place the money with an insurance company that offers him an annuity.

John's range of investment choices could not have been imagined years ago by investors, like him, of limited means. And his freedom to invest along so many potentially rewarding avenues is made possible by a wondrous, evolving capital market which, like the fecundity of Kansas wheat fields or the beauty of the Grand Canyon, greatly benefits us all.

8.

━━━━━━

Education

"Education has for its object the formation of character."
— Herbert Spencer (1820–1903), *Social Statics*

"When ign'rance enters, folly is at hand; Learning is better far than house or land."
—Oliver Goldsmith (1728–1774), *She Stoops to Conquer*

- The number of U.S. libraries has increased 13 percent, to more than 36,000, within the past decade.
- The percentage of Third World girls enrolled in primary and secondary schools has jumped to 68 percent from 38 percent.
- The more people earn, the more they continue to study.
- There are more than 15 million college students in the United States today.

A Shining Record

Remarkable progress has been made in recent decades in the field of education.

The Ford Foundation's 1995 annual report states that "in the 50 years since the end of World War II, no part of society, either in the United States or in the developing world, has undergone more rapid expansion than education." And, the report continues, "improved education is, in turn, inextricably linked to reducing poverty, enhancing individual achievement and enriching the overall economic, social and political well-being of a society."

Consider a few specifics:

- In China the literacy rate has increased from under 20 percent in 1950 to over 60 percent.

- In Indonesia, the literacy rate has climbed from only 9 percent to more than 60 percent in six decades.
- Worldwide, the percentage of children enrolled in secondary education has more than doubled since 1970, from 31 percent to more than 65 percent.
- Early in this century, only 10 percent of American eighteen-year-olds had high school diplomas, while now only some 10 percent *don't* have them.
- Higher on the educational ladder, some 42,000 Americans gain doctorates each year, up from a mere 560—yes, 560—in 1920.
- Britain now produces 160,000 university graduates a year, 50 percent more than as recently as 1990.

As such advances indicate, outlays for education have been soaring. The United States now spends nearly $300 billion on primary and secondary education, more than three-and-one-half times the 1960 level, even after allowing for inflation. Inflation adjusted spending at the college level has risen still faster, a nearly six-fold increase to $195 billion yearly. The pattern is much the same, with few exceptions, the world over. The table opposite tracks the rise of investment in education for a sampling of countries with varying degrees of prosperity.

Pakistan provides an example of how many nations are managing to boost their educational outlays by eliminating some other governmental subsidies and selling some money-losing state enterprises to private investors. One benefit of such actions is that people seeking higher education are getting more financial help. In less than twenty-five years, the major U.S. federal student financial assistance programs have grown from less than $1.6 billion to more than $25.7 billion. In the same period, voluntary financial support for college education has climbed from less than $1.8 billion to nearly $12.3 billion.

**Public Spending on
Education as Percent of GNP**

	1960	Latest
Canada	4.6*	7.6
Norway	4.6*	8.4
New Zealand	2.2*	7.1
Pakistan	1.1	2.7
Portugal	1.8*	5.0
Brazil	1.9	4.6
Gabon	2.1	5.7
Ethiopia	0.8	5.1
Zimbabwe	0.5	9.1

* Spending as a percent of GDP

SOURCE: United Nations Development Programme's *Human Development Report, 1996*

Private Aid to Education

Corporations, nonprofit organizations, and individuals also are providing more time and money to help educate the disadvantaged. The Ford Foundation, for example, is funding a $10-million project called QUASAR (for Quantitative Understanding: Amplifying Student Achievement and Reasoning). It aims to improve the mathematical capabilities of low-income sixth to eighth graders. Using models, sketches, and diagrams, students create mental images that help them comprehend and remember mathematical concepts. In Portland, Ore., a school participating in the five-year project saw the number of students qualifying to take algebra in the eighth grade jump from ten to one hundred in only two years.

Similarly, employees at Amdahl Corp., a Sunnyvale, Calif., computer maker, help teach fifth and sixth graders at

A QUASAR student in Jacqui Jolivet's class at the Magnet Middle School in Holyoke, Massachusetts, is using visual models to figure out math problems.

El Carmelo Elementary School in nearby Palo Alto, as a part of the firm's "Read to Succeed" program. Amdahl is also a corporate sponsor of First Step, a Coalition for the Homeless program that offers homeless women eight weeks of classes to train them to find and hold jobs. This is followed by internships with many of the participating sponsors, including such firms as Hugo Boss, Millennium Broadway, and Fujitsu. To date, 70 percent of First Step's graduates have landed jobs or are pursuing further education.

Individual contributions to education range from a few dollars contributed by many thousands of graduates to their alma maters to a $500 million donation to the U.S. public education system by Walter Annenberg. Actor Paul Newman uses profits from his salad dressing and food company to help support a program of after-school tutoring for minority children. And Microsoft chairman Bill Gates has given $34 million to the University of Washington and $15 million to Harvard, in addition to all his other financial contributions to support education.[1]

The Quest for More Education

There are nearly 15 million students attending colleges and universities in the United States today. That's two-and-a-half times more than thirty years ago and some one-hundred times more than a century ago. The State University of New York alone has nearly 400,000 students spread across sixty-four campuses. Some 213,000 of these students attend the City University of New York, which constitutes the largest enrollment in a single city anywhere.[2]

1. Karen W. Anderson, "Gates of Microsoft Gives $15 Million to Harvard," *New York Times,* October 30, 1996.

2. *The Guinness Book of Records 1996* (New York: Bantam Books, 1996).

In 1940, just 6 percent of all Americans in their late twenties had college educations and nearly two-thirds were high school dropouts. Today, in the same age group, around 90 percent finish high school and some 20 percent have college degrees, according to the *SAUS*. As recently as 1960, less than 40 percent of Americans twenty-five and older had completed high school. By 1995, nearly 82 percent had done so and there was nearly a three-fold increase in those holding college degrees.

Twenty years ago, less than half of high school graduates in the United States entered college within a year of receiving their diplomas, while today three in five do so. For the industrialized world as a whole, this college-entrance ratio has more than doubled in the last thirty years. Moreover, 86 percent of Americans between the ages of six and twenty-three are now enrolled in school. In Canada, the rate is 89 percent and in Spain, 83 percent. The average for industrial nations as a whole is 77 percent.[3]

Even with such numbers, student-teacher ratios are improving the world over. In Denmark, there are only ten students, on average, per teacher. Oil-rich Qatar boasts an eight-to-one ratio in its secondary schools, and a ten-to-one ratio in its elementary schools. In the United States, the student-teacher ratio in elementary and secondary schools averages about seventeen-to-one, down from more than twenty-six-to-one as recently as 1960. In Thailand, the ratio averages eighteen-to-one, which is also down sharply from several decades ago.

Nations around the world show their commitment to education in myriad ways. Japan in particular places a great deal of emphasis on the education of its citizens. More than 93 percent of all students in Japan continue their schooling at least into the upper-secondary levels, which is all the more

---

3. *Human Development Report 1995* (New York: Oxford University Press, 1995).

remarkable when one considers that Japanese schools charge fees. Japan's commitment to education is evident in the growth of *juku*, or private tutorial schools that children may attend in the afternoon and evening after their regular school day is finished. At least half of Japanese children attend these cram schools, with some 80 percent of the attendees doing so in the year before they apply for entry into a university. The *juku* system, I should add, has recently spread to South Korea and China.[4]

Today, there are twice as many institutions of higher education in the world as there were only fifty years ago. Perhaps the most remarkable growth has been in education in the field of business and commerce. Early in this century, there were only two graduate schools of business in the world, while now there are over six hundred in North America alone. After master's degrees in education itself, master's degrees in business administration are the most popular type of graduate degree in the United States, *SAUS* statistics show.

Colleges and universities are expanding in order to handle more students. Five times as many Americans earn a B.A. degree and eleven times as many go on to receive advanced degrees as was the case fifty years ago. Across the Atlantic, one in every three young Britons goes on to post-secondary education today, against only one in eight as recently as 1979. Meanwhile, enrollment in U.S. graduate schools has grown from just over one million in 1986 to around 1.2 million currently.

As such data suggest, advanced degrees are fast becoming as common as a bachelor's degree once was. Two out of three college freshmen in the United States now look toward an advanced degree, up from less than one in two as recently as 1970. Additionally, today's college students are also coming out of high school with significantly better grades than years

---

4. Hamish McRae, *The World in 2020: Power, Culture and Prosperity* (Cambridge, Mass.: Harvard Business School Press, 1994).

ago. A U.S. government report finds that 28 percent have an "A" average in high school, nearly double the comparable rate in 1970. This change is far too great to be attributed merely to "grade inflation," as some skeptics claim.

Technology Revolutionizes Learning

The learning process is greatly enhanced, of course, by the growing use of computers in classrooms. At St. Paul's School in Concord, N.H., computers are now used in most classrooms. And in the school's laboratories, students use computers to perform otherwise dangerous, prohibitively expensive, and time-consuming experiments. The school's students also use computers to engage in interactive foreign-language conversations and for extensive on-line access to libraries and museums throughout the world. "In our day, we were lucky to get a pencil," recalls a St. Paul's alumnus from the late 1940s.

In a writing class, the school's students nowadays sit on swivel chairs with coasters and can slide from their computer stations at the perimeters of the classroom to a central table and back again. A teacher typically sits at a computer console connected to each student's computer and can project any student's essay on an overhead for the class as a whole to discuss and edit.

"The medium seems to make it easier for them to give and receive criticism without hesitation and embarrassment," says the school's rector, "and it holds their attention and allows them to stay on the task longer. Now they write not for the teacher only, but for one another."[5]

Private prep schools like St. Paul's are by no means alone in applying computer technology to their classrooms. Some 98 percent of U.S. public schools are equipped with computers;

5. *The Rector's Report* and *The Record,* St. Paul's School, 1994–1995.

public schools currently use approximately 4.5 million units. That's about seven times as many as a dozen years ago, when only 78 percent of the schools had computers. In the mid-1980s, on average, 63 students were compelled to share one computer, while now there's a computer for every 11 students. In fact, spending on computer technology in American primary and secondary schools now exceeds $4 billion yearly, twice as much as outlays for textbooks and nearly double the amount spent only five years ago. Education's technology boom, moreover, isn't limited to computers. In 1992, only 14 percent of all senior high schools used interactive video discs, while now some 34 percent do. And there has been an even steeper rise in the use of CD-ROMs. The table on the next page pinpoints the various increases.

Such data augur well for educational attainment. For example, one recent study finds that interactive software can speed learning by 30 percent to 50 percent, compared with conventional methods.[6] Meanwhile, some 70 percent of teachers recently polled by the U.S. Congress' Office of Technology Assessment report that computers allow them more time with individual students; most respondents also feel that with computers their students perform better.

Computerized education is helped greatly by parents who invest in computerized home learning. Parents are buying CD-ROM-equipped computers and associated software packages at record rates. In 1995, retail sales of home-learning programs reached $572 million, according to the Software Publishers Association in Washington. That's nearly a six-fold increase since 1991, when home-learning software sales totaled $100 million. The San Francisco brokerage firm of Volpe, Welty & Co. forecasts that by the turn of the century parents will be spending $1 billion a year on software for home learning.[7]

6. Larry Armstrong with Dori Jones Yang, Alice Cuneo, and *Business Week* bureau reports, "The Learning Revolution," *Business Week,* February 28, 1994.
    7. Ibid.

*U.S. students using various technological
instruction methods, in millions*

	1992	1993	1994	1995
Interactive video discs	5.8	9.1	13.3	16.1
Modems	10.7	13.4	16.5	19.3
Networks	3.8	8.0	12.7	15.2
CD-ROMs	5.3	8.5	15.6	19.5
Satellite dishes	1.9	4.7	6.7	7.9
Cable	NA	27.3	33.5	35.8

SOURCE: SAUS

For many youths, computers have transformed the learning process into a game. Children spend hours on such programs as a Broderbund Software game in which they track down the elusive Carmen Sandiego using geographic clues. In the course of the game they learn hundreds of facts about world geography, as scenes of foreign cities come on the screen.[8] For their grandparents, the comparable learning experience was moving airplane and ship counters about on a board game featuring a flat map of the world and several key cities.

Primary and secondary school students are not the only students benefitting from computerized education, however. College students are benefitting tremendously as universities move into computers in a big way. Massachusetts Institute of Technology spends over $6 million annually on computers for undergraduates alone and Oklahoma State University spends more than $4 million a year, to give just two examples.

The American Council on Education, a Washington trade group, reports that two-thirds of U.S. colleges are increasing their computer expenditures.[9] Fledgling aeronautical

8. Ibid.
9. Melissa Lee, "Leading the Way," *Wall Street Journal*, November 13, 1995.

engineers test designs for airplane wings using computer-simulated air tunnels. Music students edit scores digitally, enhancing composing through technology. Language students take advantage of teleconferencing with students in other universities to enhance their vocabularies. Indeed, a survey of 407 colleges finds that virtually all of them—99 percent—are exploring new ways to use computers in education.[10]

Academic Systems, Inc., a private concern based in Mountain View, Calif., reports that some seven thousand students on sixteen campuses have used its mathematics-instruction software. The computerized training has helped push pass rates as much as 38 percent higher than had been achieved with conventional teaching methods.[11]

The Internet is another technological tool many students use to further their educations. Many students, for example, are using the Internet to learn about science. With it, some students are even contributing to scientific research. Grade schoolers in eight hundred classrooms from Texas to Newfoundland recently helped entomologist Orley Taylor track the path of monarch butterflies as they migrated to Mexico. The students reported sightings over the Internet, where the information was shared with other students and scientists. Similarly, some one thousand classrooms in forty-nine states and eight Canadian provinces recently followed the migration of loggerhead turtles and bald eagles, among others species, through using information supplied on the Internet by scientists.[12]

In addition, the Internet is making it easier for students to get into college. The Internet College Exchange's database provides the addresses and phone numbers of some 5,000 colleges, along with information on such things as tuition,

10. Ibid.
11. Don Clark, "Academic Systems Gets High Marks Helping to Lift Pass Rates in College," *Wall Street Journal*, April 3, 1996.
12. William M. Bulkeley, "The World's a Lab," *Wall Street Journal*, November 13, 1995.

CLOSE TO HOME JOHN McPHERSON

When teachers tele-commute.

development ahead of further rapid economic development, and a top priority is to provide nine years of basic education to every Thai child.

Distance learning is improving the quality of education not only in Thailand, but around the world. The list of other developing countries that are benefitting includes, among many others, Nicaragua, Mexico, China, and Kenya.

Rising School Enrollment in the Third World

Developing countries are also making major strides in getting students off farms and city streets and into classrooms. In developing countries, as a group, the net enrollment at primary levels has increased in the past thirty years to 77 percent from 48 percent. Over the past two decades, the percentage of Third World girls enrolled in primary and secondary schools has jumped to 68 percent from 38 percent, according to the 1995 *Human Development Report.*

As a result of such strides, major gains in literacy are evident in every region of the world. In the Arabic-speaking countries of Africa and the Middle East, the adult literacy rate has risen from 30 percent in 1970 to more than 55 percent today. In the last thirty-five years, primary and secondary enrollment has soared from 8 million young people to some 50 million. Even in sub-Saharan Africa, by and large a troubled area, adult literacy has doubled in the past two decades to 54 percent of the population and, in the last three decades, primary school enrollment has doubled and secondary school enrollment has tripled.

In Latin America and the Caribbean, enrollment at the high school and college level has increased eight-fold since 1970. College-level enrollment in developing countries in East Asia and the Pacific region has quadrupled in three decades. Students in this area are highly competitive even with the best qualified students in developed countries. In comparative math and science tests among thirteen-year-olds in twenty nations, South Korea topped the group in science with Taiwan in second place. The two countries tied for second in mathematics, behind China. An explanation for the high scores may be the large amount of time spent in school. School days in China average 251, followed by 222 in South Korea and Taiwan. By comparison, thirteen-year-olds in America spend an average of only 178 days a year in school.[17]

One of the driving forces behind Third World educational improvements is the United Nations. While best known for its peacekeeping activities and multilingual debates in the General Assembly, perhaps its greatest, though unsung, contribution to the world is in the field of education. The United Nations Development Programme (UNDP) has financed more than 500,000 fellowships in the past forty-five years. These fellowships, which range in duration from a month to several years, have made major contributions to creating leaders of developing countries. Hage Geingob, the first Namibian prime minister, for example, is a former UN fellow, and Theo-Ben Gurirab, who took over the Ministry of Foreign Affairs after Namibia won its independence, benefitted from three UN fellowships.

In Nepal, UNDP funds, on average, more than five hundred fellowships a year. Narayan Tiwari was working as a class three tax officer in Katmandu in 1972 when the government nominated him for a six-month UNDP fellowship to study tax administration at the University of California at Los Angeles. When he returned to Nepal, he wrote two university textbooks and worked his way up to become Director General of the Department of Taxation, and from there to Secretary at the Ministry of Public Works and Transport.

17. Hamish McRae, *The World in 2020: Power, Culture and Prosperity*.

Percent of University Students Abroad

Tunisia	25
Congo	28
Yemen	33
Cameroon	40
Jordan	41
Central African Republic	45
Chad	50

SOURCE: United Nations Development Programme's *Human Development Report, 1995*

In most developing nations, the search for higher education increasingly includes studying overseas, as the table above shows.

More and more American universities, at the same time, are offering students the opportunity to spend one of their four college years at an overseas campus run by the U.S. institution. And many other U.S. colleges have agreements with foreign universities allowing students to earn part of their college credits abroad. As the world becomes smaller, more Americans than ever before are studying languages of other lands. Around 1.2 million U.S. college students are enrolled in foreign language courses, nearly double the number in 1960.

Foreign language study is not limited to college students, however. The New Jersey Board of Education recently approved stiff new requirements for a high school diploma. Included in the package, to be fully phased in by 2002, is a stipulation that students must be able to speak and read a second language in order to graduate.

Continuing Education

Forty percent of all American adults participate in continuing education courses of one sort or another, including 15 percent of those over sixty-five-years-old. In Britain, meanwhile, there are more "mature" adults entering the nation's higher-education institutions than young people.

Surveys show that the more people earn, the more they continue to study. Nearly three in five Americans earning more than $75,000 annually are enrolled in an adult education course, while less than one in four earning less than $10,000 attends one.

Many educational institutions are encouraging their own employees to crack the books. Drury College in Springfield, Mo., lets its full-time employees and their dependents go to college free after the first year of employment. The benefit is good for four years of day classes and an unlimited number of

Continuing Education in the U.S. by Salary Category

Income	Total Population (millions)	Attending Adult Education Courses (millions)	Percent
Under $10,000	30.2	6.9	23
$10,001–$15,000	13.5	3.6	27
$15,001–$20,000	13.1	4.2	32
$20,001–$25,000	13.8	4.3	31
$25,001–$30,000	16.4	6.2	38
$30,001–$40,000	28.6	12.2	43
$40,001–$50,000	20.4	9.6	47
$50,001–$75,000	29.2	15.2	52
Over $75,000	24.3	14.1	58

SOURCE: SAUS 1996

night classes.[18] Massachusetts, Connecticut, and New Jersey offer free tuition at state-supported schools as a recruiting incentive for their National Guard units.

Two centuries ago, recruiting for the armed services in America was sometimes conducted by press gangs. Today, it is done through expensive persuasion. The U.S. Army offers enlistees up to $30,000 toward their college education after only four years on active duty. Another incentive is the skilled technical training that is offered on the job, as well as reimbursement for 75 percent of the cost of college classes completed during active duty.

Most large U.S. companies as well as the federal government pay part of or all the tuition costs of employees seeking to improve their skills and knowledge. Many help employees pay for any credited course, while others link reimbursement to job-related continuing education.

Additional training and education, of course, benefit the employer as well as the employee. The World Bank, for example, finds that simply teaching farmers to read has helped boost agricultural productivity 9 percent.

Educating and training employees is big business. In Britain, employers spend more than $75 million a day on training. IBM operates customer technical education centers throughout the world. In addition, it runs its own campus—the Management Development Center—next to its Armonk, N.Y., headquarters. At the campus, IBM trains hundreds of top and middle managers each year.

An estimated 480,000 managers attend the American Management Association's training seminars, conferences, and special meetings every year. The AMA, America's largest corporate training institution, reports that the total climbs to

18. Clarissa A. French, "Many Big Employers Offer Education Benefits," *Springfield (Missouri) Business Journal,* March 25, 1996.

John Cleese, 1997.

one million executives a year if you include executives receiving instruction through its books, periodicals, videos, and interactive CD-ROMs. In fact, corporate training videos have become big business, and many feature such screen personalities as British actor John Cleese.

One typical catalog promotes corporate training videos ranging from *Implementing Total Quality Management* to *How to Deal with Difficult People.* Interestingly, it also advertises a three-hour video on *Speed Reading,* a hopeful sign that books won't ever be rendered extinct by the Internet.

As we move deeper into the information age, education will become important. We surely can take great comfort in the major gains in educating people that have already taken place around the world, and even greater comfort in the clear commitment of both governments and private organizations around the world to raise educational standards to new heights.

9.

Information and Communications

"Knowledge is of two kinds. We know a subject ourselves, or we know where we can find information upon it."
—Samuel Johnson (1709–1784), *Boswell's Life of Johnson*

"Good, the more communicated, the more abundant grows."
—John Milton (1608–1674), *Paradise Lost*

- Sixteen million Americans use cellular telephones.
- Fifteen million Americans use the Internet, with 62 percent on-line for more than two hours per week.
- Electronic retail sales over the Internet are estimated to rise from $100 million in 1995 to $300 billion by the year 2000.

Astounding Advances

Some of this century's most astounding advances have come in the fields of information and communications, completely transforming the ways we are able to connect and work with one another. And, as Milton noted, raising the potential for an abundance of good. Less than fifty years ago we lived in a world without xerography, without lasers, without microchips, without man-made satellites, without fax machines, and without modems. In 1990 alone, 250 million video cassette recorders were manufactured, another item that did not exist fifty years ago.

As amazing as these inventions seemed to be when they burst upon the scene, it is even more amazing how quickly many of them are becoming obsolete. During the past twenty years, computers, telephones, and television have increased their information-carrying capacity a million times over. Computer power doubles every eighteen months, in line with Moore's Law, which is named after Gordon Moore, the co-founder of Intel. Today's $2,000 laptop computer is many times more powerful than a $10 million mainframe was in the mid-1970s. Twenty-five years ago there were only about 50,000 computers in existence. Today that number has increased 2,800 times![1]

The incredible increases in the speed and efficiency of communication have led to corresponding increases in information and knowledge. It took from the time of Christ until the mid-eighteenth century for knowledge to double. It doubled again within 150 years, and then again in only 50 years. In the last 30 years more new information has been produced than in the previous 5,000 years. Author John Naisbitt has estimated that the amount of available information would double every 20 months. At that rate, in less than 10 years, available information will have increased one thousand-fold![2]

When the Athenians defeated the Persians at Marathon in 490 B.C., the news was conveyed to Athens, some twenty miles away, by a runner. The battle of New Orleans was fought after the War of 1812 had ended, leading to thousands of lives lost because it took fifty-one days for news of the peace treaty signed in Europe to reach Washington, D.C. When Napoleon was defeated at Waterloo, Reuters scored a major news coup using carrier pigeons to dispatch the news from Europe. Yet, in the 1990s during the Gulf War a television reporter was able to sit in a besieged Baghdad, broadcasting a live commentary into American living rooms as U.S. planes bombed the enemy capital.

1. "The Hitchhiker's Guide to Cybernomics," *Economist*, September 28, 1996.
2. John Naisbitt, *Megatrends* (New York:Warner Books, 1984).

Other examples of the tremendous gains in communication technology abound. Early manuscripts, for example, were painstakingly produced by monks. In the fifteenth century, Johann Gutenberg invented printing from moveable type, which revolutionized society by making books widely available for the first time in history. In the late 1880s, a keyboard operated typesetting machine called the Linotype was invented, revolutionizing the newspaper business. Today entire pages of newspapers are transmitted via satellite to distant printing plants. Magazine pages laid out, complete with illustrations, on a personal computer are sent directly to the printing plant via a modem.

Wireless telegraph was invented in 1895, news dispatches began to move by wire instead of by messenger or pigeon. Reporters working for William Randolph Hearst's International News Service recount the tale of the legendary Jimmy Kilgallen, who single-handedly matched a three-man Associated Press bureau's output by eavesdropping on their Morse code messages through the wall between their offices with a stethoscope.

Soon teletype machines using punched tape to transmit each letter replaced the telegraph. Clicking away at sixty words per minute they printed out the news on rolls of paper, which were torn off, edited, and sent for typesetting. Today wire service news is transmitted at the speed of light, and can be edited and composed swiftly on a computer screen.

In the last century packet boats carried the mail across the oceans and along coastal regions. The Pony Express relayed bags of mail across America's western plains. But news and packages could still take weeks to reach a remote destination. By the 1920s airmail was speeding the process. Today express services offer overnight delivery of even large parcels to most areas of the world and even same-day delivery within the United States. An individual, meanwhile, can compose a letter on a laptop computer while lying on the beach or flying across the ocean and it can be faxed or e-mailed in a matter of seconds to distant parts of the globe.

A report from the Rand Corporation, meanwhile, goes so far as to laud e-mail as the new foundation for democracy around the world. The report states that "use of electronic mail is valuable for individuals, for communities, for the practice and spread of democracy, and for the general development of a viable national information infrastructure." Combined, these technological improvements have transformed work life.

In offices just forty years ago, revisions to letters and contracts meant time-consuming retyping on mechanical typewriters. Carbon paper was used to generate copies, and

"I was going to fax it to you, but that seemed so impersonal, so I'm going to e-mail it to you."

From the *Wall Street Journal*—Permission, Cartoon Features Syndicate hitting a single wrong key on the typewriter could mean as many as seven erasures.

In the 1950s, copying machines revolutionized the office world. But they were, initially, slow and messy, using a photographic process requiring specially treated paper. This was supplanted by xerography. Now, in a single business day one Xerox machine can spit out thousands of high-speed copies on plain paper, printing on both sides and even collating for magazine-sized publications.

The advent of electric typewriters meant major improvements in the appearance of the text and the speed at which it could be produced. The self-adjusting ribbon made corrections vastly easier than laboriously having to correct copy with the round eraser and brush that old-time writers and editors can still remember.

Large word processors enabled text changes to be made with minimal effort, thus speeding the ability to communicate. These processors, such as the Wang, gave way in turn to desk-top personal computers. Hooked into so-called local area or wide area networks, personal computers have, in many places, made it possible to do away with paper altogether and to transmit information instantaneously by electronic mail, or e-mail.

In the 1980s fax machines dramatically increased the speed of communication. Almost anything written, printed, or drawn on a sheet of paper can be transmitted instantly via telephone line using a fax machine. In 1996, an estimated 35 billion sheets of office paper were used in fax machines, according to a study by Business Information Systems of Boston. That's enough paper to create a ribbon that would encircle the world 241 times! Only a decade ago, by comparison, the fax paper trail would have barely made it 15 times around the globe. Ten years ago there were only about 500,000 fax machines around, and they were slow and expensive, and used almost exclusively by large, wealthy companies. But fax machines proliferated as the standard price dropped below $200 and the speed picked up to more than four pages a minute. Today, by adding a modem card to a computer, individuals can transmit and receive faxes without resorting at all to hard copy.[3]

The Power of the Telephone

When the telephone was introduced, a British leader said that while it was fine for the United States, it would not be needed in England, which was small and had many messenger boys. A little over a century later, in contrast, an eavesdropper in England was able to intercept and record a cellular phone conversation between Prince Charles and his lady friend that soon after was published half way round the world in an Australian magazine.

In the last fifty years, the percentage of households in America with telephones has doubled to 94 percent. New fiber optic lines can carry eight thousand conversations simultaneously, compared to forty-eight on the old copper wire lines. And soon a single fiber may have up to 100,000 times the present carrying capacity. This means that in only a few years one strand of fiber may be able to handle all of the calls that take place on Mother's Day in the United States, when telephone traffic traditionally reaches its peak.

Huge technological improvements have made the telephone an ever more accessible and effective tool for communications. Less than forty years ago, telephone exchanges in England were filled with long lines of operators who connected calls by hand. They timed toll calls by noting the time on a clock and recording on a special form the number of

3. Kara Blond, "Mountains of Faxes and No End in Sight," *The Record*, July 5, 1996.

minutes a customer talked. And telephones were so hard to come by that many people had to share lines with up to three neighbors. Each party had its own "ring" so you could tell when a call was for you or for your neighbor, whose conversations you could overhear just as your neighbors could listen in on you.

By the 1970s, it still took up to three months to have a telephone installed in England. Today, in contrast, this work is routinely done within two or three days, thanks in part to privatization of the system.

Moreover, calls are becoming ever cheaper as technology improves, spurring a massive increase in the use of Alexander Graham Bell's invention. Fifty years ago, it cost $12.00 to call London from the United States and talk for three minutes. It also required the services of an overseas operator. Today, such a call can be made for only 36¢. In 1945, a call between the United States and Japan cost $10.00 a minute, while today the same call costs some 95 percent less, without adjusting for inflation. In the meantime, I should add, the cost of sending an airmail letter to Japan has risen twelve-fold.

Not surprisingly, the number of overseas calls made each year from the United States has risen to more than 3 billion, more than 15 times the number made as recently as 1980 and some 4,150 times the number made in 1950, according to the *SAUS*. To help handle the surge in calls, the number of communications satellites in orbit above the earth has more than tripled to sixteen, at latest count, in the last decade.

The number of telephones in the United States has increased from 115 million in 1969 to more than 156 million today, and in most other countries the growth in telephones is even steeper. In China, the number of telephones has increased from 2.7 million to 19.5 million in the last ten years.

Now, the Internet promises to sharply reduce further the cost and vastly increase the number of overseas telephone conversations. Intel Corp. has created a free software program that enables users to make long-distance calls over the

Internet for the price of a local call. Many experts believe that in a few years telephone companies may actually abandon charges based on the number and distance of calls and charge a flat monthly fee.[4] A World Trade Organization agreement among sixty-eight countries is expected to cut the cost of an average ten-minute call from the U.S. to $2.00 from the present $9.10. The opening of former state monopolies to global competition is expected to save consumers as much as $1 trillion by the year 2010.[5]

The table on the facing page shows the proliferation of telephones in selected countries.

Telephone technology continues to improve. Personal answering machines that allow messages to be retrieved from distant locations are commonplace. It is now possible to press a simple code that will dial the number of the last incoming call and another code will continually redial a busy number. A "call waiting" option signals that a second call is coming in, enabling a person to interrupt a conversation and answer it, rather than have the telephone ring busy for the caller.

Less than forty years ago, telephone answering services were a thriving business, with operators picking up calls for doctors, salesmen, plumbers, electricians, and other entrepreneurs who were often absent from their offices or homes. Today, this has been changed due to pagers and personal answering machines with remote access. Also, most large businesses have a "voice mail" system giving callers the option of leaving a message or transferring to another individual.

Another recent development, the cellular telephone, is now used by 33 million Americans, 16 times the 1988 number. A commuter stranded in the middle of a traffic jam can call

4. Dean Takahashi, "Intel to Unveil Internet-Phone Software," *Wall Street Journal*, July 22, 1996.

5. Del Jones, "Cost of International Call Should Plunge 78%," *USA Today*, February 17, 1997.

Number of telephones in use
(in thousands):

Country	1969	1984–85	1996
Brazil	1,800	11,000	12,500
Chile	300	800	1,600
Egypt	400	1,100	2,700
India	1,100	3,200	8,500
Indonesia	200	800	1,900

SOURCES: *Statistical Abstract of the United States* and *World Almanac*

home or the office from the car to warn of a late arrival. A motorist whose car breaks down on an isolated road can call for help without any long hike to the nearest pay phone.

The growth of these devices has been phenomenal, as a few 1992 statistics indicate. In Sweden and Finland, close to 8 percent of the respective populations subscribed to cellular telephones. There were 182,000 cellular telephones in China. And in Indonesia, a country with only 2 million standard telephone lines, there were 38,000 cellular telephones. In Tashkent, Uzbekistan, technological leap-frogging occurred a couple of years ago when some enterprising Americans installed a cellular telephone system as a functional overlay to a hopelessly outdated wired system.

And now a new generation of "smart phones" is emerging. They can display the number that is calling before the telephone is picked up. Some of the latest models can even display the name and number of a second caller on "call waiting" during a conversation.

So less than eighty years after the dial telephone eliminated the need to ask a live operator to connect a call, it is possible to simply ask the telephone to call someone. For example, Bell Atlantic is offering New Jersey customers a service that lets them "train" their telephones to recognize up to fifty spoken names and connect them directly with that person.

Television Takes Off

The first network television broadcast in color was in 1954, when the Tournament of Roses parade was broadcast from Pasadena, Calif. By 1969, some 600 million viewers were tuned in to watch men walking on the moon. And just since that moonwalk, the number of television stations in the United States has nearly doubled to more than 1,500. Meanwhile, the number of television sets in the United States has increased more than 150 percent to more than 210 million. That's an average of 2.2 sets per U.S. household. And the number of cable television subscribers has skyrocketed to more than 60 million, six times the number twenty-five years ago.[6]

Television has become a powerful communications tool, with advertisers paying hefty sums for as little as thirty seconds of air time. Politicians have trained themselves to communicate with potential voters in "sound bytes" of a few seconds each. But in addition to helping advertisers and politicians communicate with the general public, television brings foreign cultures into the living room. Imported American television shows and movies are big earners for overseas television networks. Likewise, Barbara Phillips reported in a *Wall Street Journal* article that in America, "I've curled up on my living room couch . . . and watched . . . an Italian salute to mothers; Latin American telenovelas and variety shows; Greek movies; Japanese samurai epics; Indian musicals; the evening news from Moscow (and) Korean game shows."[7] In

6. *Statistical Abstract of the United States.*
7. Barbara D. Phillips, "No Passport Required, Luv," *Wall Street Journal*, July 1, 1996.

the years ahead, television is expected to become so sophisticated that a viewer will be able to call up any program or movie at any time of day or night.

In his recent book, *Being Digital*, Nicholas Negroponte of the Massachusetts Institute of Technology's Media Lab predicts that television sets will ultimately disappear, completely replaced by personal computers. "The set-top box will be a credit-card size insert that turns your PC into an electronic gateway for cable, telephone, or satellite," he writes. He also forecasts that the PC will download vast amounts of programming and information, enabling someone to create personalized TV programming or a personal newspaper bearing just information of interest to that person.[8]

Computer Power Surges Ahead

Computers, once confined to the business world and functioning largely as giant number crunchers, are now on the verge of bringing vast new changes in the way we communicate. One example is video conferencing. A new president of ITT Automotive, for example, was able to meet, face to face, with his staffs in Detroit and Frankfurt, as well as with corporate officers in New York, on the day he joined the company.

In video conferencing, a camera mounted on a personal computer sends images to other locations. A spreadsheet opened on one computer can be worked on and amended on another in a distant location. PictureTel, a video conference provider, reports that such conferences give "you most of the advantages of being together in the same room, without the cost and hassle of getting there."

8. Nicholas Negroponte, *Being Digital* (New York: Vintage Books, 1996).

Another benefit of computers comes in book publishing. Personal computers have enabled thousands of aspiring authors to write, design, and even self-publish their books. The number of new, small publishers topped 5,500 in 1995, up from less than 1,000 "self-publishers" in 1980. After rejections from eight publishers, for example, Barbara Saltzman used her computer to produce *The Jester Has Lost His Jingle*. The book was written by her twenty-two-year-son, who had died of Hodgkin's Disease. The self-published work made it to the *New York Times* best-seller list in March 1996. By April, her garage, which she had converted into a 20,000 volume book warehouse, was empty![9]

When the first functional computer—called ENIAC for electronic numerical integrator and computer—was introduced it weighed thirty tons. It contained nearly eighteen thousand vacuum tubes to enable it to multiply fourteen ten-digit numbers in a single second. At its recent fiftieth birthday celebration, Vice President Al Gore said: "Because of that clunky old machine . . . the way we work has changed. The way we think has changed."[10]

And, of course, computers have changed dramatically as well. All of ENIAC's power now can be reproduced on a modern microchip measuring less than one-tenth of a square inch! Moreover, scientists today are on the verge of building computers 100 billion times faster than ENIAC. Already computers can multiply 1.8 trillion ten-digit numbers in the time it took ENIAC to multiply fourteen![11]

At present, it is possible to buy a greeting card that, when opened, delivers a ten-second personal birthday message. This novelty card contains more processing power than

9. Doreen Carvajal, "Do-It-Yourselfers Carve Out A Piece of the Publishing Pie," *New York Times,* April 28, 1996.

10. Matt Crenson, "New Day of Reckoning Dawned 50 Years Ago," *Dallas Morning News,* published in *The Record,* January 19, 1996.

11. Ibid.

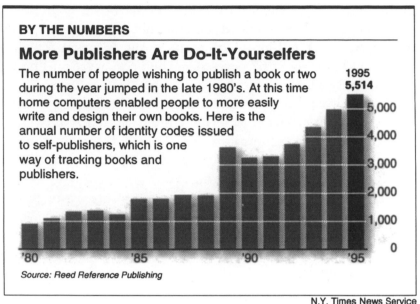

BY THE NUMBERS

More Publishers Are Do-It-Yourselfers

The number of people wishing to publish a book or two during the year jumped in the late 1980's. At this time home computers enabled people to more easily write and design their own books. Here is the annual number of identity codes issued to self-publishers, which is one way of tracking books and publishers.

1995
5,514

5,000
4,000
3,000
2,000
1,000
0

'80   '85   '90   '95

Source: Reed Reference Publishing

N.Y. Times News Service

the combined power of all the vacuum-tube computers that existed in 1950.[12]

The Amazing Internet

Gates notes that virtually all Microsoft's software efforts now revolve around the Internet. Until the early 1990s the Internet was almost exclusively used by government agencies and universities that could access it with mainframe computers. Today, with the Internet thrown open to anyone with a personal computer and telephone line, the global computer network is growing at the rate of 85 percent a year. A 1996 study

12. Don Peppers and Martha Rogers, "As Products Get Smarter," *Forbes ASAP,* February 26, 1996.

by Computer Intelligence Infocorp., of La Jolla, Calif., estimated that there were 15 million Internet users in the United
States. Some 11 million of them used the Internet for electronic mail and about 62 percent of the users were on-line for
more than two hours a week.[13]

The Nordic countries and the United States are leading
the way in Internet access. The United States has twenty-
four Internet host computers per thousand people, slightly
ahead of Norway's twenty-two, but well behind world-leader
Finland's forty-two per thousand.

The Internet allows individuals to tap into a vast storehouse of knowledge from their living rooms. The World
Wide Web connects users to more than fifty thousand databases. And these sites, or "home pages," are increasing at
the rate of one thousand a day! There are Web sites devoted
to every imaginable subject, from river rafting and basketball to the works of Shakespeare and Proust, making the
World Wide Web a massive unedited encyclopedia—the
biggest in history.[14]

The move is on to make this incredible access to knowledge available to all. Several cable companies are offering
free Internet access to U.S. elementary and secondary schools
via high-speed cable modems. The cable connection will provide dramatically faster access to the global computer network than traditional telephone lines. The National Cable
Television Association expects more than three thousand
schools in sixty communities to be wired in the first year of
the plan.[15] At the same time, the New York Public Library is
among the many American libraries that offer Internet access

13. "About 15 Million Users Troll the Internet, A New Study Finds," *Wall Street Journal,* May 23, 1996.

14. David Wallechinsky, "Be at Home on the Internet," *Parade Magazine,* November 19, 1995.

15. "Cable Companies to Offer Internet Access to Schools," *Wall Street Journal,*
July 10, 1996.

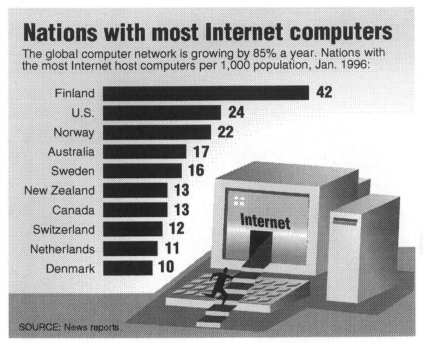

Nations with most Internet computers

The global computer network is growing by 85% a year. Nations with the most Internet host computers per 1,000 population, Jan. 1996:

Finland	42
U.S.	24
Norway	22
Australia	17
Sweden	16
New Zealand	13
Canada	13
Switzerland	12
Netherlands	11
Denmark	10

SOURCE: News reports

2/28/96   Knight-Ridder Tribune/TIM GOHEEN

to their patrons. Its $9.5 million system enables the general public to use its 221 personal computers to locate articles from 2,600 magazines and journals. Of these, 1,000 offer full texts on line.[16]

The Internet also offers tremendous new opportunities for disseminating product information. Large corporations protect their market shares using expensive advertising campaigns in print and on radio and television. The makers of Budweiser beer, for example, spend millions of dollars for a handful of television spots on a single Super Bowl Sunday.

16. Carey Goldberg, "Moving Light-Years from Dewey Decimal," *New York Times,* November 14, 1995.

Patrons use computers. New York City Public Library.

But a small company can now reach consumers throughout the world by setting up a Web page for less than $1,000.

Microsoft's Bill Gates predicts that when the Internet reaches its full potential, possibly within the next decade, it will become the world's central department store, allowing the smallest entrepreneurs to reach markets in every corner of the globe. Going on-line eliminates many of the traditional barriers to entering a national or global market, such as marketing and retail distribution. Perry and Monica Lopez, for example, turned their three-hundred square-foot Hot Hot Hot store in Pasadena, Calif., into a global company by using the Internet. Now they sell their hot sauce to customers in Switzerland, New Zealand, Brazil, and other distant locations.[17]

Amazon.Com Inc., an on-line bookstore, is able to offer more than one million titles on the Internet. Its chief executive says that "you can't do that with a physical store, and you can't do that with a paper catalog, which would be the size of seven Manhattan phone books." And, of course, with lower overheads he can discount more titles than his more traditional rivals.[18]

Imagine the comprehensive information the Internet can provide the comparison shopper! The Computer Intelligence survey finds that some 2.7 million Americans use the Internet to shop or obtain other commercial services. Jerry Berry, a senior analyst at the company, thinks consumer buying "will increase gradually over the next two to three years, and in about five years we'll see an explosion in people using the Internet to buy and sell services."[19]

Market researchers Killen & Associates estimate that electronic retail sales over the Internet will increase three thousand times in the next five years, rising from $100

17. Jim Carlton, "Think Big," *Wall Street Journal,* June 17, 1996.
18. Jared Sandberg, "Making the Sale," *Wall Street Journal,* June 17, 1996.
19. "About 15 Million Users Troll the Internet."

Irresistible Appeal?

Projected on-line shopping revenue (in millions)

	1996	1997	1998	1999	2000
Computer products	$140	$ 323	$ 701	$1,228	$2,105
Travel	126	276	572	961	1,579
Entertainment	85	194	420	733	1,250
Apparel	46	89	163	234	322
Gifts and flowers	45	103	222	386	658
Food and drink	39	78	149	227	336
Other	37	75	144	221	329
TOTAL	$518	$1,138	$2,371	$3,990	$6,579

Source: Forrester Research, Inc.

million in 1995 to $300 billion by the year 2000.[20] Another more conservative estimate is given by Forrester Research, Inc., which expects on-line shopping revenue to jump from $518 million in 1996 to $6.6 billion in the year 2000.[21] Above is a breakdown by product groups.

In his book, *The Road Ahead,* Bill Gates predicts that a wallet-size personal computer will replace money, credit cards, address books, appointment books, keys, cameras, cellular telephones, fax machines, and pagers.[22] So the second sort of knowledge that Dr. Johnson spoke of—retrieved information—may soon be almost as easily accessed as what's already within our heads.

20. Michael Moynihan, *The Coming American Renaissance* (New York: Simon & Schuster, 1996).

21. Jared Sandberg, "Making the Sale."

22. Bill Gates, *The Road Ahead* (New York: Penguin, 1995).

10.

███████

Transportation

- American automobile manufacturers now produce between 5.4 and 8.2 million vehicles a year.
- The number of U.S. accident fatalities dropped 25 percent between 1980 and 1994—despite the steep increase in the number of vehicles on U.S. streets and highways.
- The number of passenger complaints against U.S. airlines fell from 40,998 in 1987 to 4,629 in 1995.

Decades of Remarkable Progress

Progress in the movement of goods and people has been remarkable over the past one hundred years and truly astounding in the last half century. When I was born most Americans had never been more than one hundred miles from their birthplaces.

By 1925, a quarter century after the first autos jiggled along, a U.S. highway system of 200,000 miles was under way. By a recent count, there are about 1 million miles of federal highway, a six-fold increase from the mid-1920s. Today in America, millions of cars, buses, and trucks sweep daily over 2.3 million miles of paved streets and highways, up from 1.6 million miles in 1940 and 204,000 miles in 1904, according to the Federal Highway Administration.

Before the advent of the automobile, however, remarkable advances in rail transportation were being achieved. As far back as 1830, trains began to replace horses and, by the second half of the nineteenth century, the railroads enjoyed an "unchallenged supremacy" in the field of transportation around the world.[1]

Although trucks are gaining on them, railroads remain the major U.S. freight haulers. At the same time, cars and planes dominate passenger travel, as the accompanying table shows. The chart on the facing page shows how the various modes stack up in the United States. The figures are in billions of ton-miles (one ton hauled a mile) and in billions of passenger-miles (one person moved a mile).

The Auto's Ascendancy

Driving comfortably along a smooth four-lane highway while listening to a soothing symphony on a CD player implanted in the car's dashboard, a motorist today may find it hard to believe that drivers of an earlier era worried constantly about engine stalls and tire blowouts, never dreaming that one day these problems would be a thing of the past and that unknown entities like on-board sound systems would be de rigueur.

The largely trouble-free cars of today are no longer a luxury. To meet demand, production has soared from a minuscule 4,195 in 1900 to an annual output of between 5.4 million and 8.2 million since 1975, according to the Automobile Manufacturers Association.

Despite the steep rise in the number of autos over the years, the number of deaths due to auto accidents is dropping. In 1994, for example, there were 40,676 fatalities in motor

1. Paul Hastings, *Railroads: An International History* (New York: Praeger Publishers, 1972).

Modes of Transport in the U.S., 1970–1995

	1970	1980	1990	1995
All Freight	1,936	2,487	2,895	3,384
Railroads	771	932	1,091	1,375
Truck	412	555	735	921
Water	319	407	475	476
Oil Pipelines	431	588	584	599
Domestic Airways	3	5	10	13
All Passenger	1,181	1,467	2,034	2,364
Private Cars	1,026	1,210	1,639	1,906
Domestic Airways	119	219	359	415
Bus (a)	25	27	23	29
Railroads (b)	11	11	13	14

(a) Excludes school and urban buses.
(b) Includes Amtrak and commuter service.
SOURCE: *Eno Transportation Foundation*

vehicle accidents in the United States. That's more than 25 percent below the comparable 1980 figure.

The public's growing passion for autos is evident in figures showing that tire sales growth far exceeds population growth. Increasingly, it appears, each family wants more than just one car to get around.

Airplanes Take Off

Airline passengers who routinely fly between continents at speeds of more than five hundred miles an hour may find it hard to relate to this fact: less than a century ago, a man in an engine-powered, winged contraption flew for twelve seconds about ten feet above a North Carolina beach.

Commercial daytime flights in bouncy biplanes, with pilots watching railroad tracks below for guidance, didn't have

Auto Registrations in the U.S., 1900–1995

	Auto Registrations	Percent Gain	Population	Percent Gain
1900	8,000		75,994,575	
1910	458,377	5,629	91,972,266	21
1920	8,131,522	1,673	105,710,620	14
1930	23,034,753	183	123,202,624	16
1940	27,372,397	18	131,669,275	6
1950	40,190,632	46	150,697,361	14
1960	61,430,862	53	179,323,175	19
1970	89,279,800	44	203,302,031	13
1980	121,601,000	36	226,542,199	11
1990	143,453,000	18	248,709,873	9
1995	191,071,000	33	263,835,000	6

Note: Percentage gain is from end of prior ten-year period. The 1995 total is estimated and includes vans and utility vehicles.

SOURCES: Census Bureau; U.S. Department of Commerce; U.S. Department of Transportation

their faltering starts until 1918 and remained trouble-ridden through the 1920s. The few pilots who dared to fly at night followed the sight of smudge pots and beacon lights below. It wasn't until late 1932 that the first regular night service across the United States began, with onboard radios following ground-based tower beams, beacon-to-beacon. Today, by comparison, planes are beginning to use global satellites to ascertain their aerial positions.

In 1954, the airlines really took off as they entered the jet era with the intercontinental Boeing 707. With ocean liner traffic declining already, the mid-1950s saw more U.S. passengers flying to Europe than sailing there. By then, moreover, three times as many passengers traveled around the United States on planes than trains. Since 1970, the

wide-bodied, 500-passenger Boeing 747 jetliner has acceler-
ated the trend to air travel. And, of course, the supersonic
Concorde, winging along at incredible speeds of more than
1,320 mph, has added glamour to air travel, albeit at fares
that relatively few people can afford.

Below is the statistical record of the growth of U.S. air
travel over the decades. The table shows the number of passengers boarding planes and the total miles they have flown
in domestic and international travel.

Despite this growth, the number of complaints by passengers against airlines has fallen from 40,998 in 1987 to
4,629 in 1995, the latest year available, according to the U.S.
Department of Transportation. What a remarkable decline!

A Bright Future Lies Ahead

If the approaching century comes close to equaling the one
near an end, humankind can safely expect still more

Plane Travel, 1926–1995

	Passenger Boardings (thousands)	Passenger Miles (millions)
1926	6	1
1940	2,966	1,152
1950	19,220	10,243
1960	57,872	38,863
1970	169,922	131,710
1980	296,903	255,192
1990	465,560	457,926
1995	547,384	540,399

SOURCE: Air Transportation Association

11.

Leisure

"Increased means and increased leisure are the civilizers of man."
—Benjamin Disraeli, *Speech in Manchester*, April 3, 1872

- U.S. sales of sporting goods have more than tripled since 1980, an appreciably steeper climb than the increase in the size of the entire American economy during the same period.
- American consumers spend approximately $20 billion a year on books, up from $10 billion as recently as the early 1980s.
- Leisure time, and the activities that go along with it—travel, eating out, reading, participating in sports, going to museums, plays, and movies—are increasing worldwide.
- Tourism is the world's largest employer and the world's number one export industry in terms of revenue.

The Fruits of Labor

Among the many blessings that flow from a rising standard of living is increased leisure time. Over the years, the number of hours that may be spent in leisurely pursuits, rather than at work trying to earn a living, has increased greatly for most Americans, as well as for most people in other major industrial nations.

In America—and the American experience is by no means unique—the length of the average work week has diminished from some fifty-three hours at the start of the twentieth century to about thirty-five hours at present. There has been a five-hour reduction in per-week working time just since the end of World War II. The length of the average work *year* has grown shorter as well, thanks in large measure to the greater availability of paid vacations, more special holidays and sick days, and more time off for such matters as maternity—and paternity—leave. In fact, the average factory worker in the United States now works nearly one hundred hours a year less than his predecessor did as recently as 1973.

Elsewhere around the world, statistics show a similar story. The work year has shortened on average by nearly three hundred hours in the same span, and the average work week has declined by around 15 percent just since 1977. In Finland, some 16 percent of total labor costs, including social security taxes and fringe benefits, as well as salaries, are for time not worked. In Sweden, this tab comes to 12 percent and in Portugal, 11 percent.

Much of the reason for Finland's high rate can be explained by its generous annual leave, or vacation, allowances. The average manufacturing employee in this Nordic nation gets on average a whopping thirty-eight days of vacation each year—some seven-and-a-half weeks. Though this is more than three times as much as the comparable U.S. average of twelve days, it barely exceeds the thirty-two days averaged in Italy and the Netherlands, the twenty-seven days averaged in Sweden, and the twenty-five days averaged in the United Kingdom and France.

Bear in mind that these are merely averages. In France, where the major cities traditionally "shut down" for August, some newspaper workers receive ten weeks paid vacation. Adding in 104 days for weekends, plus another 10 days for national holidays, these journalists are able to enjoy leisure activities for 45 percent of the year on full pay!

You may question whether maternal and paternal leave really should qualify even partly as leisure time, but no one can doubt that such practices make life much easier for new parents than it once was. In Germany, a mother can remain away from work until a new child is three, and then get her job back. In Austria, the leave period is two years. In Belgium, parents get a total of five years off to mind children they produce.

Not only is there more time than ever before in history for leisure activities during our working lives, but today people are living longer—and remaining active longer after retirement—than ever before.

Since 1900, average life expectancy in the United States has risen to seventy-four years from forty-seven, a gain of nearly 60 percent. In the early years after World War II, barely 8 percent of the U.S. population was over sixty-five years old, while the comparable rate today is 13 percent. Looking ahead, people over the age of sixty-five will make up about 20 percent of the U.S. population—some 100 million men and women—by the middle of the twenty-first century. Already, there are more than 30 million Americans older than sixty-five.

A major benefit of this increasing longevity is that "retirement is no longer just a transient afterpiece, the brief interregnum before death," observes Michael Norman, a journalism professor at New York University. People have "achieved something of a second summer," he adds, the first phase of which "represents the time from 65 to 85—from retirement to the onset of physical decline."[1]

A century ago, in bleak contrast, a typical retiree in the United States, if fortunate, might linger on for perhaps as long as a decade after packing up work. But a more typical "interregnum" was only three or four or five years. And this brief retirement period was usually far less active and less pleasurable.

Today's seventy-five-year-old retiree may well be found on the golf course, the tennis court, or at the gym in the

1. Michael Norman, "Living Too Long," *New York Times Magazine*, January 14, 1996.

morning and at the movie house or perhaps the concert hall in the evening. The scene for such a septuagenarian decades ago would likelier have shifted back and forth between the sitting room and the bedroom, with perhaps a slow walk around the block thrown in on sunny days.

Their increased leisure time has benefited Americans in countless ways. It improves the quality of life and grants new opportunities for pleasure and learning to millions of people whose forebears knew little else away from home but on-the-job drudgery. Indeed, many economists argue convincingly that, on account of increasing leisure, the U.S. economy in reality is appreciably stronger than customary measures show. As one economic textbook puts it, growth in the gross national product "systematically *understates* the growth in national well-being" since the data largely fail to take improving leisure time into account.[2] (This, of course, directly contradicts the misperception, widely circulated in the press and elsewhere, that GNP data systematically overstate national well-being.)

For all of this good news, however, concern still is voiced that all too often this extra time is merely wasted in idleness. A frequent complaint is that television has produced a generation of "couch potatoes," people who squander their spare hours glued to the TV set, vacantly watching tawdry talk shows, soap operas, and the like.

Enriched Lives

In fact, people by and large are utilizing their increasing leisure hours to enrich their lives. Consider this statistic: sales of sporting goods in the United States have more than

2. William J. Baumol and Alan S. Blinder, *Economics: Principles and Policy,* 5th ed. (New York: Harcourt Brace Javonovic, 1991).

tripled since 1980, an appreciably steeper climb than the increase in the size of the entire U.S. economy in the same period. The variety of items for which demand has surged—and continues to rise apace—ranges from exercise machines and snowmobiles to fishing tackle and aerobic shoes. This hardly suggests a nation of sit-at-homes. Americans now spend, for instance, nearly $3 billion a year on exercise equipment, some $10 billion on athletic clothing of various sorts, $3.7 billion on bicycles, and $1.4 billion on golf equipment.

Nor does this vast array of athletic gear languish unused in attics and closets. A survey of U.S. households by the National Sporting Goods Association found that some 35 million Americans work out with exercise equipment in the course of a year, nearly 60 million engage in fresh or salt water fishing, 23 million play golf, 40 million go on hikes, jog, or run for exercise, some 11 million do calisthenics, 15 million ski downhill or cross country, more than 41 million bowl, 61 million swim, 25 million engage in aerobics, 48 million ride bicycles, more than 50 million backpack or camp out, 65 million engage in exercise walking, and many millions more regularly target shoot or play tennis, volleyball, baseball, basketball, and racquetball. Participation in such activities, I should add, is up sharply from years ago. As recently as 1970, only some 9 million Americans played golf, for example, and fewer than 8 million bowled.

Along with this surge in sports participation has come a steep increase in attendance at sporting events. More than ever, Americans are using their leisure time to be sports spectators, rather than simply watching sports by way of television. Since 1985, attendance at major league baseball games has risen from less than 48 million to more than 71 million. In other big-league sports, the story is much the same. In the same period, attendance at National Basketball Association games has climbed from 11.5 million to about 20 million; at National Football League games, from 14 million to nearly 18 million; at National Hockey League games from 11.6 million to more than 16 million.

Engaging in sports, or attending sporting events, is but one of many stimulating ways in which Americans are enjoying their leisure hours. For all the talk of TV couch potatoes, the truth is that U.S. consumers are shelling out record amounts of income for books. As recently as the early 1980s, such spending came to slightly less than $10 billion annually, or about half the recent yearly rate. Spurring this increase, moreover, has been surging demand for hardbound books dealing with serious topics, such as college texts, books on religion, and books that treat issues in various occupations. Book prices have been on the rise of late, but even so Americans are buying ever more books. Yearly purchases now exceed 2 billion books, up from about 1.7 billion in the middle 1980s. It should be noted, in addition, that purchases of hardcover books have risen appreciably faster than purchases of softcover books, even though the latter tend to be much less expensive and often deal with frothier topics.

Some book stores, in fact, have become leisure palaces in their own right, catering to the increased interest and spending by the public. Barnes & Noble, a U.S. book store chain, provides coffee bars and cushy easy chairs for browsers to relax in, plus large play areas for children. The chain even invites scholars and authors to give free in-store lectures.

Travel is another way in which Americans are gaining greater enjoyment from their leisure time. Two hundred years ago, the swiftest means of transportation—by horse or perhaps by sailing vessel—were hardly conducive to vacation jaunts to remote locations. Now, jetting to Europe for a week of sight-seeing fun or education is as routine for a vacationing family from the midwest as once was a week at the Jersey shore for a vacationing family from Manhattan.

International travel is growing in both directions, of course. In 1988, Newark International Airport received and

The Barnes & Noble store in Metarie, Los Angeles, California stocks more than 175,000 different titles and features a popular in-store cafe.

dispatched slightly more than one million international passengers, flying on only two airlines. Now, less than a decade
later, over 3.6 million international passengers pass through
its terminals annually en route to or from forty foreign cities
on twenty-five airlines. For perspective, only forty years ago
this airport was a base of the U.S. Air force, commanded by a
brigadier general!

In 1950, some 25 million tourists throughout the world
ventured abroad. By 1983, this number had rocketed to 292
million, and today the figure is well in excess of 500 million. In
1950, receipts from international tourism totaled $2.1 billion.
Today, international tourists spend about 150 times that much.

Antonio Savignac, a former secretary general of the
World Tourism Organization, a United Nations agency, estimates that by 1995 tourism had displaced oil as the world's

number one export industry, in terms of revenue. It is also, by far, the world's largest industry in terms of employment. The World Travel and Tourism Council estimates that one in every nine jobs worldwide is generated by tourism. By 2006, the London-based group estimates, travel and tourism will account for 385 million jobs, or one in every eight. This represents an increase of 50 percent in employment in a single decade, or one new job every 2.4 seconds!

The incredible boom in tourism is being fueled mainly by rising incomes and increased leisure time in industrialized nations. In turn, this boom has created an unprecedented number of new jobs and opportunities for people in developing countries. For example, the number of tourists visiting the tropical beaches and Buddhist temples of Thailand has increased to around 6 million, up from 4.8 million as recently as 1989. International tourism is now the country's premier foreign-exchange earner. Thailand is feverishly building condominiums, hotels, and convention centers for foreign visitors and pressing marketing campaigns to capture an ever increasing slice of the global tourist pie, which itself is expanding. The country also is creating exercise centers for health-conscious foreigners and retirement villages for western pensioners seeking lower prices and less crowded living conditions than may be available at home.

Elsewhere in Asia, the tourism boom continues unabated. India, where tourism is the fastest growing sector of the economy, expects to double hotel occupancy in the course of the 1990s, a feat that is requiring the nation to train some 250,000 new workers, from waiters to hotel managers. In 1988, Indonesia entertained 1.3 million tourists, most of them at the island paradise of Bali, and the number has since jumped to more than 4 million. This surge has created hundreds of thousands of new jobs. Similarly, the Caribbean Tourist Organization estimates that tourism, which is the island region's largest industry, provides income for some

560,000 workers. Even Castro's Cuba is hoping that vacationing capitalists will make tourism that country's leading business by the end of this century.

Helping to spur tourism at home, of course, is the proliferation of the automobile. In the early 1920s, only about one
U.S. family in four was fortunate enough to own a car. By the
advent of World War II, this ratio had risen—the Great Depression notwithstanding—to more than one in two and it
now exceeds nine in ten. Today, moreover, one family in two
owns more than just a single vehicle and, remarkably, one
family in five owns more than two.

With such mobility, it is no surprise that travel to such
remote areas as those managed by the U.S. Forest Service, for
instance, has grown increasingly popular. Estimates by the
Forest Service show that Americans now spend some 300 million "recreation visitor-days" each year relaxing in these often
remotely situated forests. That's a record amount of time and
represents an increase of some 70 million visitor-days just
since 1980. (A visitor-day is the recreational use of these
areas for twelve visitor-hours—that is, a visit by one person
for twelve hours, twelve people for one hour each, or any
equivalent combination.) Particularly sharp is the rise in visitor-hours spent camping—up nearly three-fold in only a
decade. Yosemite Valley, for example, has become such a popular place to spend leisure time that for the first time recently
some visitors have had to be turned away on busy summer
weekends.

The pleasure that Americans derive through outdoor
recreation nowadays is greatly facilitated, I should add, by a
huge expansion in the amount of acreage that has been set
aside in recent decades for leisure time use in the United
States. The size of the National Park System, which covered
3.3 million acres in 1900, now encompasses nearly 80 million
acres. Similarly, the amount of acreage maintained by the
various state park systems has more than doubled to nearly

20 million acres just since World War II. And the amount of land within the National Wildlife Refuge System has approximately quadrupled to about 80 million since 1950.

Not every form of outdoor recreation, to be sure, is growing apace. There has been a modest reduction, for example, in family picnicking. More than offsetting such declines, however, is swift growth in such long-standing outdoor activities as bicycling, walking for exercise, and swimming. In addition, a range of relatively new and increasingly popular outdoor activities has been made possible by technological advances. The list includes sail boarding, jet water-skiing, snorkeling, hang gliding, and, for the bravest of the brave, bungee jumping.

Varied Entertainment

Americans are also seeking more entertainment during their leisure time, and much of it involves stepping out, rather than sitting at home. Outlays to attend motion pictures now approximate $6.5 billion annually, up nearly $2 billion in half a decade, and outlays to attend concerts, theatrical productions, and the like total about $14 billion, a rise of nearly $6 billion in only five years.

This is not to say that Americans are forsaking entertainment within the home. Some 63 percent of U.S. households are now equipped with cable television, more than triple the rate in 1980, and about 80 percent have VCRs, four times the 1980 rate. Thus, the choice of home-entertainment fare is vastly greater nowadays. There are all sorts of specialized channels and video tapes, such as for golfers, investors in the stock and bond markets, weather buffs, home shoppers, students of various languages, close followers of Washington political events—and the list goes on.

Television, to be sure, has long been a fixture in U.S. households; nearly 99 percent of all homes have TV today,

Bobby Bronze recoils after taking his jump at Bungee Canada's Vancouver location.

which is only slightly higher than the rate as long ago as 1970. However, today's sets are much improved and there are more of them per household. In 1970, less than 7 percent of U.S. homes had color sets, for example, while two-thirds do now. And in 1970, the average number of sets per home was 1.4, while today's average is 2.2.

The table below provides a panoramic snapshot of the increasing amounts that U.S. consumers are spending for recreational pursuits. The dollar amounts are in billions, expressed in terms of the dollar's 1992 buying power to adjust for inflation, and in most cases the latest year available is 1994.

Young Americans, like their senior counterparts, are active participants in this recreational surge. There are more than 25 million teenagers in America today and, looking ahead, the nation's teenage population is expected to expand

Amount Spent on Recreational Pursuits by U.S. Consumers

	1970	Latest
Total Recreation Outlays	93.8	369.9
Percent of Total Consumer Spending	4.3	8.3
Video & audio products, computer equipment, musical instruments	6.2	89.0
Books and maps	12.8	19.1
Magazines, papers, sheet music	16.7	22.5
Wheel goods, photo equipment	11.7	38.2
Radio & television repair	3.2	4.4
Flowers, seeds, potted plants	4.8	14.0
Spectator amusements	10.9	18.3
Clubs, fraternal organizations	4.5	11.5
Pari-mutual net receipts	3.8	3.0
Pets, video rentals, cable TV, etc.	14.2	79.3

SOURCE: U.S. Bureau of Economic Analysis, *Statistical Abstract of the United States, 1996*

in the next decade at nearly twice the rate for the U.S. population as a whole. These younger Americans already spend more than $90 billion a year, about two-thirds of which is from their own earnings and the rest from allowances. Outlays by eighteen- to nineteen-year-olds, for instance, now average about $80 a week, up steeply from $30 a week as recently as 1987. Much of this money goes for items involving outdoor recreation and entertainment. Teenagers spend, on average, some $3 weekly on athletic shoes, for example, and fully 60 percent of the nation's twelve- to seventeen-year-olds have their own CD players; Sony Corp. estimates that one in five belongs to a record club. Indeed, teenagers account for nearly 25 percent of record sales in the United States.

Other leisure time items that are highly popular with teenagers nowadays include portable phones, video games, and VCRs. This a vastly more interesting assortment, I need hardly note, than what was available for teenagers decades ago. In the early years after World War II, for instance, a sixteen-year-old who enjoyed listening to records had to manage with large, awkward, fragile discs that required frequent changing and utterly lacked today's quality of sound. And, of course, there were only a handful of television programs for children. Today young people, with proper supervision, can learn about the sciences, the arts, and current events by watching programs created especially for them. For example, the early childhood program *Sesame Street* has helped a generation of children learn their ABCs. Television programmers have also created award winning after-school programs for early adolescents. And, of course, the personal computer—such an integral part of the youth culture today—simply did not yet exist.

Computer Culture

Advancing technology, in fact, has played a huge role in improving the quality of leisure time for people in general.

Consider the advent of what is frequently referred to as computer culture. Institutions ranging from New York City's Bettmann Archive, which houses the world's largest collection of historical photographs, to the Andy Warhol Museum in Pittsburgh, to the Museum of Paleontology in Berkeley, Calif., now make at least parts of their displays available to PC-users via the Internet or by way of CD-ROMs.

As a result, people wishing to view, say, a particular painting by Matisse or a tintype of a black Union infantry corporal in the Civil War need not travel to a distant museum or gallery. Rather, thanks to technology, they now can simply feed the appropriate commands to their PCs and—presto!—enjoy the art without having to step outside the home. Through zoom-lensing, moreover, technology now permits close-up scrutiny of, say, the brushstokes in a particular masterpiece.

Making such a computer culture possible are such individuals as Bill Gates, Microsoft's legendary chairman. He recently purchased the remarkable Bettmann Archive of some 16 million images, which he is digitalizing and licensing for use on-line via computers and CD-ROMs. Through his Corbis Corp., Gates also provides computerized access, for example, to some 330 works housed at the Barnes Foundation in Merion, Pa., by such artists as Matisse, Renoir, and Cezanne. This computerized access also provides written text and voice-over commentary about the particular work and artist. This may not quite match the thrill of actually seeing such art at a museum or gallery, but for many Americans it provides a welcome alternative to not seeing the art at all.

Though the variety of cultural entertainment available at home is vastly greater today than years ago, families continue to increase their visits to such places as museums. Americans now spend more than $200 million yearly at museums and botanical and zoological gardens, roughly twice as much as in 1990. Among museums that have recently expanded to accommodate larger crowds are two New York City institutions—the Museum of Modern Art and the Guggenheim Museum.

At the same time, the number of cultural associations in the United States is on the rise, with the total recently reaching a record 1,904. There has been a similar proliferation of other sorts of associations that also serve to make Americans' leisure hours more interesting. The number of national nonprofit groups devoted to hobbies recently reached 1,555, up more than 70 percent since the early 1980s. Similar gains are evident among groups that engage in sporting events, scientific discussions, and religious activities.

A Widening Trend

The trend toward a more leisurely, more pleasant, and more interesting life—so evident in the United States—is taking hold in other regions of the world as well. In Asia, for instance, people by and large have more money to spend and more time in which to spend it than ever before. In Japan, where workers once toiled far longer hours than their U.S. counterparts, the situation is now reversed. The average Japanese worker now labors an average of 1,800 hours a year, some 250 hours *less* than the comparable rate in America. "Not only do Asians have money to spend, they have more time in which to spend it," observes John Naisbitt, who conducted a recent survey of the region. Asian "attitudes toward work and leisure," he declares, "are changing."[3]

A telling sign that many Asians are beginning to lead more pleasurable lives is that more of them are occasionally indulging in a treat once reserved for only the very rich—eating out. In China, McDonald's has recently opened twelve outlets in Beijing alone, with four of the openings occurring in a

3. John Naisbitt, *Megatrends Asia* (New York: Simon & Schuster, 1996).

single weekend. The restaurant chain's plans call for more than six hundred outlets in China by 2003, with one hundred of them in Beijing. Already, there are more than one thousand McDonald's in Japan. Kentucky Fried Chicken, Kenny Rogers Roaster, and T. G. I. Friday's are other restaurant chains that are in the process of rapidly expanding their Asian operations.

Asians are also stepping out more often to fancier places. A 16,000-square-foot version of Dallas's famous Hard Rock Cafe recently opened with considerable success in Beijing, and the restaurant and entertainment chain also operates popular franchises in Kuala Lumpur, Jakarta, Taipei, Bali, and Bangkok.

Asians are also using their increased leisure to travel more for pleasure. "In this part of the world, there are tens of millions of people in the middle class starting to travel for leisure," says Gerald Pelisson, co-chairman of Accor SA of France. Accor plans to build 123 hotels throughout the Asia-Pacific region, he reports, including 20 in Vietnam and 5 in Malaysia. A sign of the times is that more Chinese tourists visited Malaysia in a recent twelve months than did American tourists.[4]

Until recently, only a tiny number of wealthy Asians were able to enjoy vacation cruises, but now a rapidly growing segment of the population is undertaking such excursions. A $10 million luxury vessel, the *East King*, now cruises the Yangtze River laden with vacationing Chinese families. It is equipped with bars, a gymnasium, a sauna, and a telephone in each of its seventy-eight staterooms.

Leisure hours have also become a lot more pleasurable nowadays for Asians who choose to spend their spare time at home. Some 400 million Asian homes now have TV, and in many nations—Japan, Singapore, Taiwan, and South Korea, among others—the percentage of households with sets is

4. Ibid.

*Ronald McDonald added another language, speaking Cantonese to two
children at the opening of the first McDonald's in China.*

about as high as in the United States. The choice of TV fare, moreover, has expanded greatly in the last few years. Indians, for example, may now select from no less than sixty cable channels, supplied by some fifty thousand private cable-TV operators. Elsewhere, by a recent count, there are twenty-five cable channels in South Korea and as many as one hundred in Singapore. And in Thailand every household "now has a receiver for home entertainment—whether a TV, a radio, VCR, stereo or CD player," declares Paiboon Damrongchaitam, who heads a diversified supplier of entertainment to the region.[5]

In sum, people everywhere are enjoying more leisure time, which in turn is helping to enrich their lives. This blessed trend is yet another reason for us to be so very thankful that we live in such a wonderful time in history.

5. Ibid.

12.

▬▬▬

The Environment

"Our searches . . . forced us to revise our earlier, naive view that environmental deterioration has been a universal, accelerating process."
—William Baumol and Wallace Oates, *The State of Humanity,* 1995

- Overall American air pollution levels are down to one-third of the levels that existed on the first Earth Day in 1970.
- The rain forests are being deforested at the rate of only 0.08 percent per year.
- From 1988 to 1993, the U.S. recycling rate rose from 3 percent to 22 percent.
- The amount of carbon byproducts expelled into the planet's atmosphere is decreasing even as worldwide energy use increases.

The Truth Versus the News

News stories about the environment tend to be overly pessimistic in the extreme. They warn of global warming, a growing hole in the earth's ozone layer, species extinction, destruction of tropical rain forests, ruined rivers, and devastating oil spills along pristine coastlines. The true outlook, however, is very different. In fact, all the gloomy news about the environment obscures a growing body of evidence that points in precisely the opposite direction—to real progress in making the environment cleaner and healthier.

Though serious attention to preserving the environment began less than thirty years ago, the major industrial countries already are turning the corner on solving their environmental problems. New technologies for preventing and remediating environmental damage are being developed and put into use, and powerful economic incentives are spurring this progress.

Hand-wringing about the environment, however fashionable it may be, largely ignores such developments, as well as the fact that the earth has long been subject to environmental affronts. As long ago as the second century A.D., Tertullian wrote, "What most frequently meets our view is our teeming population; our numbers are burdening to the world."

If we could transport ourselves back in time, we would encounter environmental problems that would make our current challenges look easy. In colonial Massachusetts, entire forests were felled and fields disrupted by the plow. Industrial facilities in the late 1700s routinely dumped toxic waste into streams. Great rivers such as the Thames, the Hudson, and the Cuyahoga became open sewers from which fish disappeared. Until the 1930s, Londoners choked—many to death—on dense smog from coal furnaces and coal fireplaces. Between 1945 and the passage of the Limited Test Ban Treaty in 1963, more than four hundred nuclear weapons were detonated in the atmosphere, spreading radioactive substances through the air. It is no wonder that Rachel Carson, writing in the 1960s, could correctly state that a "battery of poisons of truly extraordinary powers" was being poured into the environment as pesticides, killing birds, fish, and other animal life.

Today, remarkably, such lamentable situations have been eliminated or largely corrected. For example:

- There are now more acres of forest in Massachusetts than at any time since the American Revolution.
- In most industrial countries, waterways are protected from industrial and household pollution to the point that fish and other wildlife are returning to once-devastated rivers.
- London's killer smogs are now a historical footnote.

- There is an international agreement to ban nuclear bomb tests in the atmosphere.
- The highly toxic DDT pesticide that inspired Rachel Carson's landmark book, *Silent Spring,* has been replaced by ones that are environmentally benign.

Seeking evidence of long-term changes in environmental quality, William Baumol and Wallace Oates, widely respected authorities on the subject, concluded recently, "Our searches ... turned up evidence that ... broad statements reported in the popular press were often either misleadingly simplistic or completely untrue."[1]

This chapter pinpoints the progress being made in improving our environment—in air and water quality, in the condition of the world's forests, and in other ways.

Air Quality

Back in 1970, the year of the first Earth Day, the typical U.S. automobile burned one gallon of gasoline to drive just twelve or so miles. Engine emissions went directly into the atmosphere. And since most gasolines had lead additives, these cars produced plenty of lead emissions along with carbon monoxide and other pollutants. Likewise, the emissions of coal-burning power plants, industrial furnaces, and refineries were largely unregulated. In places like Sudbury, Ontario, emissions from metal smelters devastated foliage and lakes for dozens of miles downwind.

Today, overall levels of air pollution are down to about one-third of those that prevailed on Earth Day 1970, as the following figure shows. And this has occurred even though

1. William J. Baumol and Wallace E. Oates, "Long-Run Trends in Environmental Quality," *The State of Humanity*, Julian L. Simon, ed. (Cambridge, Mass.: Blackwell, 1995).

Air Pollution Levels, 1970–1991

	Sulfur Dioxide	Nitrogen Oxides	Carbon Monoxide	Total Particulates	Volatile Organic Compounds	Lead
1970	28.42	18.96	123.6	23	27.4	0.2
1980	25.51	23.58	99.97	10	21.75	0.06
1991	20.73	18.76	62.1	5	16.88	0.005

(in million metric tons)

SOURCE: *Council on Environmental Quality Annual Report, 1992*

many more cars are driving many more miles, and much more electricity is being generated from coal-burning power plants.

Data from Canada and the United Kingdom parallel the United States data.[2] One reason for this dramatic improvement is that automobiles coming off the assembly line today emit just 1 percent of the pollution per mile that 1970 models emitted—an astounding improvement.[3] At the same time, emissions of chlorofluorocarbon (CFC), the prime suspect in ozone layer depletion, are near zero in major industrialized nations thanks to an international agreement and the development of a benign substitute for CFC.

Even regions where air quality has been especially poor show a notable improvement. Southern California, for example, still has the dirtiest air by far of any large area in the United States. Even so, the region's air recently reached its cleanest level in the forty years in which records have been kept. In a recent May-through-October period, which is the re-

---

2. For Canadian air pollution trends, see *The State of Canada's Environment, 1991*. (Ottawa: Ministry of Supply and Services, 1991).

3. Gregg Easterbrook, "Good News From Planet Earth," *USA Weekend*, April 14–16, 1996. For full details, see Easterbrook, *A Moment On The Earth* (New York: Viking Books, 1996).

gion's so-called smog season, the number of high-level smog alerts totaled only seven. This represented a precipitous reduction of some 90 percent from the comparable alert level in the mid-1980s and was far under the record of 121 alerts, set in 1977.[4] This remarkable showing is widely attributed to reduced emissions from motor vehicles, which in turn reflects the introduction of unleaded gasoline and the adoption of catalytic converters. Also helping has been the recent implementation in California of a new type of gasoline that emits less exhaust. The change equates to removing 3.5 million cars from California's roads, according to the State Air Resources Board.

Water Quality

The quality and quantity of water in streams, rivers, lakes, ponds, and aquifers is obviously important to households and businesses, as well as to wildlife and vacationers. Water, of course, is subject to degradation from many sources: pesticides and fertilizers that run off from agricultural land; chemical and fuel spills; inadequate sewage treatment and more. Notwithstanding, it is evident that water quality in the United States has improved greatly in recent years, and further improvement seems likely.

In 1970, only 25 percent of river-miles in the nation met federal standards for fishing and swimming, while now, amazingly, 60 percent meet that standard. Boston Harbor, for example, was considered practically "dead" in 1970, a victim of poor sewage treatment and industrial pollution in feeder rivers and streams. Today, in happy contrast, water quality in the harbor has improved to the extent that lobstermen work there and area beaches have reopened for swimming. All this

---

4. "Still Worst in U.S., California Air Is at Cleanest Level in 40 Years," *New York Times*, October 31, 1996.

has happened, moreover, even though most of the harbor's scheduled water-treatment projects are not due to come on line for several more years. Like Boston Harbor, Lake Erie, Chesapeake Bay, and Long Island Sound also were near death in 1970, but are now in much improved health.

One of the great success stories in U.S. water quality is the venerable Hudson River. In 1966, a *New York Times* story titled "Life Abandoning Polluted Hudson River" described an appalling scene: diseased fish encrusted with something resembling cottage cheese; cities and towns discharging untreated sewage directly into the river; and bottom sediments loaded with toxic industrial waste. Swimmers, boaters, and fishermen had abandoned the river.

Since that story was published, however, two forces have combined to restore the Hudson's vitality. The first was a long-term change in the economy of the river valley itself. Agricultural activity in the valley diminished, bringing a reduction in the run-off of agricultural chemicals from adjacent farms. At the same time, antiquated manufacturing facilities gave way to businesses that were more environmentally benign, including such service industries as tourism. As old factories closed, the level of chemical waste entering the river dropped sharply. The second major force was the federal Clean Water Act of 1972, which brought major spending on waste treatment plants up and down the Hudson and crackdowns on major polluters.

In the past two and a half decades, these measures and trends have significantly reduced pollution to the river. The *New York Times* revisited the Hudson in 1996 and found that the swimmers, boaters, and recreational fishermen were back. Water oxygen levels at many locations on the river were at twice 1960 levels. Harbor seals had reappeared and the striped bass were back in record numbers.[5] In brief, the Hud-

5. William K. Stevens, "Shaking Off Man's Taint, Hudson Pulses With Life," *New York Times*, June 9, 1996.

son River had regained its central place in the life of the natural and human cultures of the river valley.

In the Hudson and elsewhere, the dumping of PCBs (polychlorinated biphenyls), an industrial compound for use in electrical components, was outlawed in the mid-1970s. And despite fears that PCBs would remain at high levels for a century or more after their use was curtailed, indications are that their levels have begun to drop sharply, which is a happy surprise.

If water is becoming cleaner in the United States, it is also being used more wisely. For instance, the United States now bans toilets with five-gallon flushes in favor of ones that use much less water per flush. As the figure below indicates, per capita water use roughly quadrupled between 1900 and 1970. Since then, however, usage has dropped at a rate of 1.3 percent per year. Water use by industry has dropped especially steeply, as the costs of water treatment become more widely recognized.

Total U.S. Per Capita Daily Water Use, 1900–1995 (Gallons)

SOURCE: *Council on Environmental Quality Annual Report, 1992*

Land and Forests

A great deal of media attention has been paid in the past decade to the loss of tropical rain forest acreage, particularly in the Amazon basin and parts of Asia. Such deforestation can lead to the extinction of many species of plants and animals, the loss of potential sources of future medical drugs, and, with the loss of forest acreage that serves to absorb carbon dioxide, an increase in greenhouse gases in the atmosphere. But researchers have found that deforestation of the rain forests is occurring at a rate of only 0.08 percent per year, a snail's pace. Meanwhile, the amount of forest acreage outside the tropical rain forests has either remained stable or increased.[6]

In the United States, it is encouraging to recall, the early settlers stripped away vast forest areas, but by 1905 forest acreage began to increase again and continues to do so.[7] The same pattern is likely to emerge in the rain forests. Deforestation in the tropics is driven by the fact that local governments treat such land as a "free good" to be exploited at very low cost. This makes the clearing of land for lumber and marginal grazing land—commodities in abundant supply around the world—economically viable. However, as the rising value of these lands is recognized, clearing will become uneconomic and diminish. In fact, this is already happening in some places. In Costa Rica, Ecuador, and Peru, for example, rain forests are being preserved because two economic forces make it smart to do so: tourism and pharmaceutical research.

The latter bears special mention. Currently, several major pharmaceutical companies—including Inverni della Beffa, Merck, and Eli Lilly—are involved in ventures that

6. Roger A. Sedjo and Marion Clawson, "Global Forests Revisited," *The State of Humanity*, Julian L. Simon, ed. (Cambridge, Mass.: Blackwell, 1995).
7. Ibid.

seek new drugs in rain forest biota in return for royalties and cash payment.[8] Under an agreement in 1991 with Costa Rica's National Institute of Biodiversity, for instance, Merck consented to pay the Institute $1 million to set up conservation programs in exchange for access to the nation's insects, microbes, and plant life. The company also agreed to let the institute share in any royalties from medicines that it may develop as a result of the arrangement.

Pollution Prevention

Efforts to prevent pollution continue to expand. Curbside waste recycling containers have become commonplace in major urban areas. Recycled waste, in turn, precludes the use of materials that would otherwise be taken from nature through mining, logging, and so on. Recycling also reduces the need to set aside space for landfills. Between 1988 and 1993, the number of urban recycling programs grew from 600 to 3,700 in the United States, a massive gain. Thanks in part to these programs, the U.S. rate of materials recycling rose in the period to 22 percent from only 3 percent.[9] Since 1990, recycling has accounted for more than half the volume of metal consumed in the United States, almost double the rate in the 1960s.

Less visible, but more profound in their beneficial impact, are the many programs being developed by private industry to reduce pollution *at its source*. This is accomplished through either "closed-loop" manufacturing processes that recapture waste materials and energy or by "designing out" pollution from new products and the processes that create them. Some of these programs are spurred by the need to comply

8. Thomas Carr, Heather Pedersen, and Sunder Ramaswamy, "Rain Forest Entrepreneurs," *Environment,* September 1993.

9. Gregg Easterbrook, "Good News From Planet Earth."

with more rigorous environmental regulations, but others are driven by a more basic motivation: pollution prevention reduces business costs and can create competitive advantages.

As the maker of the ubiquitous Pentium computer chip, Intel Corp. is in the forefront of the information age. The information industry is lauded for its low impact on the environment, but the processes used to make its key subcomponents require the use of dozens of hazardous chemicals and prodigious amounts of water. So when Intel built a $1.6 billion state-of-the-art chip plant in Chandler, Ariz., in 1995, it designed it with environmental challenges in mind. An estimated five million gallons of water were needed to operate the plant each day. But to cut down this water requirement, Intel did some serious innovating. It contracted to buy effluent from the town's water treatment plant and recycled it through its own cooling towers. Some of the recycled water is used for landscape irrigation. The rest is used directly in the plant's manufacturing processes, after which it is purified through reverse osmosis. This water is so clean that it can be pumped directly into the natural underground aquifer from which the town draws its water supply. This process has also reduced the plant's water requirements by 80 percent.[10]

Minnesota-based 3M Corporation, maker of Scotch tapes, Scotchgard fabric protector, Post-It notes, and many other consumer and industrial products, provides another example of how companies are helping their shareholders and improving the environment at the same time. Where possible, 3M designs its products and manufacturing processes in ways that reduce the volume of pollutants created, instead of spending money to clean up pollution after the fact.

In 1975, 3M launched its 3P program (Pollution Prevention Pays) to find ways to eliminate pollution. Since then,

10. John H. Cushman, Jr., "U.S. Seeking Options on Pollution Rules," *New York Times*, May 27, 1996.

some 4,200 pollution solutions have been implemented. For example, a resin spray facility had been producing some half million pounds of "overspray" each year, which required special incineration disposal. New equipment costing $45,000 eliminated most of the overspray, reducing 3M's resin costs by $125,000 each year.

With its 3P program, 3M has cut its pollution per unit of production in half, preventing more than 600,000 tons of pollutants and 2.7 billion gallons of waste water from ever being generated. This effort, in turn, has saved 3M an estimated $750 million in operating costs.[11]

Looking ahead, 3M's research lab in St. Paul is developing ways to make existing 3M products without use of hazardous solvents. Solventless processes for making tape and adhesives will also require less energy and ultimately cost less to run.

It is useful to recall, in this regard, that perhaps the worst industrial pollution in recent decades occurred in nations that were part of the former Soviet bloc. Their economies were mainly guided by Communist planners with little concern about holding down production costs and no shareholders to worry about. But now, as these same nations turn increasingly toward capitalism and the economics of a highly competitive marketplace, pollution prevention is receiving a high priority since to do so can help reduce costs in the long run.

A Welcome Trend in Energy

Perhaps the most common environmental worry for people as they look ahead is that accelerating economic development and population growth in the Third World will place unprecedented strains on the environment. What will happen as the

11. Environmental Technology and Services Department, 3M Corporation, 1994.

multitudes in such developing countries as China and India acquire automobiles and other big-ticket consumer items already commonplace in highly industrialized nations? Citizens of the United States and Canada, for example, use energy to support their lifestyles to the extent that some twenty tons of carbon dioxide is given off per person per year. That compares with only about three tons per person in developing countries.[12] Can the earth's atmosphere survive if billions of Third World residents increase their carbon dioxide output to North American levels?

As worrisome as this question may seem, innovation has a way of solving such matters before they become too acute. For one thing, the developing economies are not starting off, as they industrialize, with the highly polluting smoke-stack factories of the early Industrial Age. Rather, they largely utilize state-of-the-art production facilities, such as those springing up in China.

Their methods of energy production, moreover, are a far cry from the polluting techniques that were common only a few decades ago. Writing in an issue of *Daedalus,* a publication dedicated to the environment, Nebojsa Nakicenovic points to the "decarbonization" of energy as a potential solution to current concern over economic growth and world atmospheric pollution. He notes that large secular decreases in energy requirements per unit of economic output have been achieved throughout the world, as we have learned better how to make, operate, and use energy systems. Furthermore, the emissions of carbon dioxide from energy systems, coming from the combustion of the carbon molecules that wood, coal, oil, and gas all contain, have also decreased per unit of energy consumed. This *decarbonization* of the energy system proves to be emblematic of its entire evolution.[13]

12. Martin W. Holdgate, "The Environment of Tomorrow," *Environment,* July–August, 1991.
13. Nebojsa Nakicenovic, "Freeing Energy From Carbon," *Daedalus,* Summer 1996.

Primary Energy Intensity, including biomass, per unit of value added from 1855 to 1990, expressed in kilogram of oil equivalent energy per constant GDP in 1990 U.S. dollars (kgoe/US$ 1990).

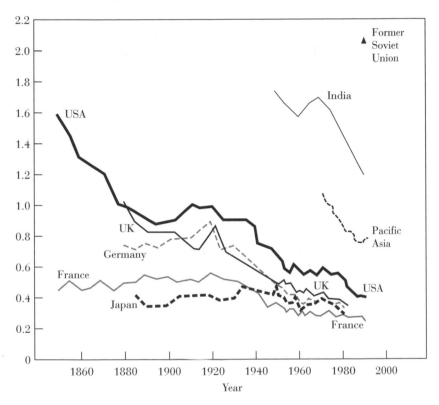

SOURCE: Nebojsa Nakicenovic, "Freeing Energy From Carbon," *Daedalus,* Summer 1996.

As the figure above shows, India today is at approximately the level of energy intensity per unit of value that the United States was at a century ago. That leaves much room for improvement, to be sure, but as the chart also shows giant strides are in fact being made.

While the energy required to produce a unit of economic value is dropping, the carbon content of each energy unit is

Carbon Intensity of Global Energy Consumption, expressed in tons of carbon per ton of oil equivalent energy (tC/toe).

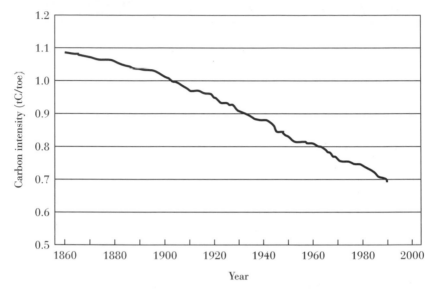

SOURCE : Nebojsa Nakicenovic, "Freeing Energy From Carbon," *Daedalus,* Summer 1996

also diminishing, as the figure above shows. In Nakicenovic's calculation, the ratio of carbon emissions per unit of primary energy consumed has decreased at an annual rate of 0.3 percent since 1860.

The news that the carbon content of energy is diminishing is enormously important in terms of the environment. Humankind has progressively shifted its major sourcing of energy from high-carbon fuels, such as wood and coal, to fuels with less carbon and greater hydrogen content, namely to oil and natural gas (methane). Each shift has reduced the level of carbon emissions per unit of energy.

Consider the chemical composition of these fuels. Wood has ten carbon atoms for every hydrogen atom and coal has roughly one for one. Oil has only one carbon atom, on average, for every two hydrogen atoms. Methane, by comparison, contains only one carbon atom for every four hydrogen atoms. Energy production by nuclear, wind, solar, and hydropower means, of course, zero carbon, and, of course, these are renewable energy sources.

The very good news here is that the amount of carbon byproducts thrown into the atmosphere is dropping even as levels of energy use rise. Gregg Easterbrook provides a preview of this energy-clean future in his description of a new natural gas power plant built in the town of Doswell, Va. The facility burns about a third less fuel per kilowatt than comparable power stations and causes minimal pollution:

> A visit to the Doswell power station is an eerie experience, for even when the turbines are running at full throttle it is difficult to tell that the facility is operating. . . . No clouds of anything rise away, not even steam. Inside the control room are no dramatic banks of color-coded warning lights and sirens, just a few desktop PCs. The plant makes 656 megawatts of power, nearly as much as one of the Three Mile Island reactors.[14]

Plants such as the Doswell generating station represent the beginning of the end for highly polluting approaches to energy generation and for large-scale use of fossil fuels. Indeed, the future is already here.

Growing Cooperation

When the Apollo astronauts first set foot on the moon in 1969, they observed the earth as no humans had before them: a bright blue planet, dappled with white clouds, spinning slowly in the cold emptiness of space. Viewed from afar, the

14. Gregg Easterbrook, *A Moment On The Earth.*

planet represented a single ecosystem with a single atmosphere that recognized no artificial boundaries of nationhood. One astronaut reflected on how the earth seemed like a spaceship whose crew all breathed the same air and lived off the same life-support system.

In the decades since then, we have all come to understand that environmental issues have no national boundaries, and that all nations must work together toward continued environmental health and progress. To that end, a growing number of multinational institutions and agreements have been created to set environmental standards, exchange knowledge and technology, and regulate harmful behavior. There are now more than 170 environmental treaties in force, and the number is soaring.

Listed on the right are several of particular significance. As with international trade agreements, each nation sees a clear benefit in collaborative efforts aimed at maintaining environmental well-being.

International Environment Initiatives and Agreements

- *Convention on International Trade in Endangered Species.*
- *The Global Environment Facility.* This facility provides financial and other assistance to developing countries for innovative environmental projects that provide global benefits.
- *The U.S.-Canada Air Quality Accord.* This agreement set standards to measure and try to reduce cross-border air pollution.
- *The Montreal Protocol.* This landmark treaty regulates emissions of ozone-depleting chemicals and provides for a complete phase out of most of them, including CFC, by 1996. The Protocol allows developing countries additional time in complying with the treaty.
- *The "Earth Summit."* Sponsored by the UN, this conference on Environment and Development in Rio de Janeiro was attended by the heads of 192 nations and thousands of scientists, industry leaders, and private citizens. The conference led to the Rio Declaration, a nonbinding statement that sets forth 27 basic principles to guide environmental protection and economic development worldwide.
- *Intergovernmental Panel on Climate Change.* This panel was created by the UN's World Meteorological Organization to monitor climate change and possible responses.

13.

Getting Along

". . . And they shall beat their swords into plowshares, and their spears into pruning hooks: nation shall not lift up sword against nation, neither shall they learn war any more."
—Isaiah 2:4

"Ring out the thousand wars of old,
Ring in the thousand years of peace."
—Alfred Lord Tennyson (1809–1892), *In Memoriam, A.H.H.*

- Worldwide military spending was reduced by $935 billion between 1987 and 1994.
- Active duty military personnel are decreasing in number all around the world.
- Many countries now spend more on health and education combined than on the military.

Former Foes, New Allies

People and nations are getting along better than ever before. Nations are sharply reducing military spending and magnanimously contributing billions of dollars toward relief and rehabilitation. Former foes have become allies in the quest for peace. People are learning to respect and get along with each other as never before.

Peace is breaking out all over. Just think of all the amazing and wonderful strides toward peace that have taken place in just the last dozen years. After nearly three-quarters of a century, Communist rule ended in what until recently was the Soviet Union. The Cold War, which had gripped the world for nearly fifty years, ended without a shot being fired. In 1989, the Berlin Wall was torn down and Germany was re-united as a democratic regime. Elsewhere, South Africa at last acceded to decades of world pressure to end apartheid. Free elections were held, with people of all races allowed to vote for the first time in April 1994. Earlier, a 1988 agreement between South Africa, Cuba, and Angola put an end to years of guerrilla warfare in the former territory of South-West Africa. Following the withdrawal of foreign forces and a United Nations-supervised election, Namibia became an in-dependent state in 1991. El Salvador, Nicaragua, Mozam-bique, Angola, Ethiopia, Eritrea, and Cambodia are among other nations recently moving from internal warfare toward peaceful stability. Even seemingly intransigent foes in the Middle East have moved gingerly from the battlefield to the negotiating table. Likewise, top-level negotiations have also served to curb the carnage in Northern Ireland.

Stockpiles of nuclear weapons have been dwindling since the end of the Cold War. And, since a single nuclear bomb contains more explosive power than all the explosives used in all the wars ever fought, world leaders have come around to the view that all-out war would be suicidal and therefore senseless.

Indeed, all-out war is no longer considered a viable solu-tion to international conflicts. This attitude can be seen, among other places, in the increasingly cooperative efforts of various nations to maintain world peace under the aegis of the United Nations.

The Foundations of Peace

The many encouraging trends that I have mentioned in other chapters generally make peace and harmony more likely to prevail in the years ahead.

The remarkable spread of democratic forms of government around the world clearly helps, since democracies almost never opt to fight one another. Major wars in the past were launched mostly by totalitarian leaders seeking to expand their realms.

The information revolution also contributes to our getting along better with one another. In fact, the Dalai Lama has confidently predicted that a solid century of peace lies ahead. He bases his optimism on the fact that nowadays global information is instantly accessible. This makes it impossible, he believes, for aggressive national leaders to insulate citizens from news from abroad and employ jingoistic propaganda to mobilize war efforts against other nations.[1] For example, the unfettered flow of ideas and information through electronic media, which the Soviet government could not control, surely played a major role in the collapse of Communism there.

The worldwide improvement in living standards, noted in an earlier chapter, is another major development that serves to help us get along better with one another. Indeed, Julian L. Simon, the economist, predicts that "within a century or two all nations and most of humanity will be at or above today's Western living standards." Since healthier, better fed, better educated people are less likely resort to war to improve their circumstances, Simon maintains that "the economic reasons for war diminish."[2]

The great growth of multinational corporations also helps contribute to a more peaceful world. These companies obviously want to safeguard their far-flung investments, which could be put at risk if war erupts. Moreover, the multinationals bring managers from around the world to work together, which helps promote tolerance and understanding among people of different cultures and nationalities.

---

1. Claudia Dreifus, "Peace Prevails," *New York Times Magazine*, September 29, 1996.

2. Julian L. Simon, *The State of Humanity* (Cambridge, Mass.: Blackwell, 1995).

Dismantling the Arsenals

Indicative of a more peaceful world is the dramatic drop in defense expenditures in recent years. Between 1987 and 1994, the cumulative reduction in military spending around the world came to around $935 billion, which is the rough equivalent of a year's global military outlays. "There is growing recognition that security in an interdependent world requires cooperation, not confrontation, and that economic vitality and environmental stability are more important to a country's fortunes than martial qualities," claims Michael Renner of the Worldwatch Institute.[3]

The drop in defense spending is not limited to the superpowers. Reductions by the United States and the nations of the former Soviet Union account for most of the estimated decline, but most other nations are following suit. Industrial nations as a whole, including the United States and Russia, reduced spending 23 percent, adjusting for inflation, between 1985 and 1994. In the same period, developing countries cut defense spending 22 percent, again adjusting for inflation. Even countries in West Asia and the troubled Mideast cut their military outlays, with particularly steep declines in Iran and Iraq. Saudi Arabia, noted in the past for its heavy spending on defense, showed a 41 percent drop, from $23.6 billion to $13.9 billion.

Defense Industry Consolidations

The global military industrial complex is undergoing unprecedented consolidation. In the United States, for instance, firms like Rockwell International are spinning off their

3. Michael Renner, "Budgeting for Disarmament," *Worldwatch Paper 122*.

Defense Expenditures
(Billions of U.S. dollars—1993 Prices)

Country	1985	1994
Argentina	4.8	3.3
China	26.1	27.7
Germany	46.3	34.8
India	8.2	7.3
Indonesia	3.1	2.3
Iran	18.7	2.2
Iraq	16.9	2.6
Israel	6.6	6.5
Saudi Arabia	23.6	13.9
Soviet Union	230.0*	110.0**
Syria	4.6	2.4
United Kingdom	41.9	33.9
United States	339.2	278.7
All developing countries	179.7	140.0
All countries	1,027.4***	778.1

*Includes estimate for former Soviet Union.
**Baltic and CIS countries previously in Soviet Union.
***Includes estimate for former Soviet Union and Czechoslovakia.
SOURCE: *Human Development Report 1996,* Oxford University Press

defense operations. The Pentagon, concerned about paying huge sums for unused capacity following the end of the Cold War, has encouraged the trend. Since 1993, twenty-two major mergers and acquisitions involving defense production have taken place between U.S. companies. A recent case in point is the $13.3 billion merger of Boeing and McDonnell Douglas.

In Russia, with sharply lower national arms requirements and diminishing arms exports, the military industry is shrinking sharply. A top Russian priority is to convert its vast military facilities and technologies to civilian use. Western governments and aid organizations are pouring in millions of dollars to support this effort. For example, with assistance from the United Nations Development Programme, a plant formerly used to produce jet engines for airplanes and rockets is being converted to the production of combined heat and power units. Market research is being conducted to test demand in Russia and elsewhere for the ten-to-twelve megawatt units, and joint-venture partners are being sought.

China, too, has been switching a good deal of its arms industry to produce civilian goods. In 1991–95, the country invested the equivalent of about $720 million for this purpose.[4]

Besides downsizing their defense industries, many nations are slashing the size of their armies. In 1970, the United States had 3.1 million military personnel on active duty, with more than 1 million stationed in foreign countries. By 1988, the active duty roster had dropped to 2.1 million, with 541,000 serving abroad. Now, according to the Statistical Abstract of the United States, only 1.6 million are on active duty, and only 267,000 serve in foreign countries.

When Russia undertook its huge, peaceful military withdrawal in Central and Eastern Europe in the early 1990s, more than 1.2 million soldiers and civilian dependents headed home. Meanwhile, Germany made grants and loans totaling some $7.6 billion to help pay for the exodus of Russian troops from the former East Germany.[5]

After Belarus declared its independence from the former Soviet Union, it undertook a large-scale disarmament process. Dozens of military bases were closed and their nuclear missiles dismantled. Some 100,000 demobilized military personnel and their families have opted to continue living

4. Ibid.
5. Ibid.

within the residential areas of some three hundred abandoned bases. Western governments and aid agencies are actively trying to find productive work in the private sector for these former soldiers.

Less for Arms, More for Health and Education

These steep military cutbacks have allowed governments to spend much more on health, education, and other important social services. Indonesia, for example, now spends twice as much on health and education as it does on the military. That's a complete turnaround from 1960, when it spent far more on defense than on health and education. Likewise, Portugal now spends three times as much on health and education as in 1960, when these sectors combined received far less than the military.

In 1960, military spending by all nations was slightly higher than the amount spent educating people and caring for their health. Today, in contrast, nations spend nearly three times as much on these social services as on military needs. And, since the most recent data is for 1991, the turnaround is undoubtedly even more pronounced than the figures indicate. Developing countries alone now spend two-thirds more on education and health than on the military, while thirty-five years ago they spent 43 percent more on the military than on education and health.

Military spending in developing nations has traditionally focused on keeping rulers in power. But this pattern appears to be changing. Costa Rica, for example, abolished its army in 1949 and created new institutions to provide its citizens with more help in education, health, and other social needs. Major industrial nations, meanwhile, have been pouring billions of

Active Duty Military Personnel (Per 1,000 Population)

COUNTRY	1986	1994
China	3.9	2.5
Cuba	29.5	12.9
East Germany	14.4	
Egypt	9.5	7.0
France	10.2	8.7
Germany (unified)		4.5
Greece	20.7	19.9
Iran	7.3	8.4
Iraq	49.9	21.4
Israel	47.8	36.6
Nicaragua	23.4	3.4
North Korea	38.5	52.0
Russia		9.3
South Korea	14.6	16.6
Soviet Union	16.1	
Syria	38.8	21.5
United Kingdom	5.9	4.4
United States	9.6	5.8
Vietnam	16.5	11.7
West Germany	8.1	

SOURCE: *World Almanac,* 1997 and 1989

dollars into other nations to help them recover from war and internal strife.

At a meeting in Tokyo in 1992, for example, the major industrial nations pledged $880 million for reconstruction and peace-building in Cambodia. This was nearly $300 million more than the secretary-general of the UN had requested. The commitment followed the Paris Peace Agreement ending

Military Expenditure
(As percent of combined education and health
expenditures)

Country	1960	1991
Argentina	62	51
Canada	66	15
China	387	114
Ecuador	104	26
France	131	29
Greece	145	71
Haiti	100	30
Indonesia	207	49
Jordan	464	138
Pakistan	393	129
Portugal	156	32
Russian Federation	134	132
Thailand	96	71
Turkey	153	87
United States	173	46
All developing countries	143	60
World	104	37

SOURCE: *Human Development Report 1996,* Oxford University Press

some thirty years of war, genocide, and foreign occupation that had shredded Cambodia's social fabric and devastated its economy. The undertaking involved 22,000 UN personnel, 17,550 military personnel, 3,600 civilian police, and 850 consultants. The UN's Transitional Authority took over most government functions and registered more than 4.7 million Cambodians to vote, or about 96 percent of those eligible. The project also helped in the repatriation of some 365,000 refugees, mainly from Thailand, and assisted another 80,000

displaced persons in returning to their original homes. The
UN also oversaw a free election in which fully 90 percent of
eligible voters cast ballots. It also began demining operations
and repairs to the nation's infrastructure.

Around the world, the amount that governments spend
annually on demilitarization and peace building has steadily
climbed. The outlays increased more than six-fold in a recent
six-year period, reaching nearly $16 billion in the latest year
for which statistics are available.

In 1992–96, the United Nations Development Programme, the UN's major funding organization, gave grant
assistance of nearly $67 million to the countries of Eastern
Europe and those formed by the breakup of the Soviet Union.
UNDP also raised many millions more for its programs from
various donor nations. The European Union is another major
donor, contributing billions of dollars to humanitarian aid

Global Peace and Demilitarization Expenditures
(Millions of U.S. dollars)

Category	'89	'90	'91	'92	'93	'94
Demining	10	10	197	200	238	241
Demobilization	46	28	38	54	56	52
Repatriating Refugees	77	101	160	172	252	463
Disarmament	1,505	1,634	3,641	4,473	5,495	5,343
Base Closures	N/A	538	998	1,148	2,120	2,864
Conversion	93	114	511	1,302	1,609	2,707
Peacekeeping/ building	749	677	760	2,149	3,450	4,080
World Court/ War Crimes	6	9	9	9	9	20
Total	2,486	3,111	6,314	9,507	13,229	15,770

Source: *State of the World, 1995*

around the world. In addition, the International Monetary Fund and World Bank are offering tens of billions of dollars in loans to newly democratic nations.

Dismantling Nuclear Weapons

In the early 1990s, two major nuclear disarmament agreements were signed by Russia and the United States. The first agreement, START I, calls for the reduction of offensive arms by about 30 percent in three phases over seven years. In the second, START II, Presidents George Bush and Boris Yeltsin agreed to do away entirely with land-based multiple-warhead missiles.

Since these agreements, moreover, the United Nations Conference on Disarmament has helped negotiate the Comprehensive Test Ban Treaty. The treaty, which bans further testing of nuclear arms, was confirmed by the UN General Assembly in November 1996 and signed by the United States, Russia, and more than one hundred other UN member states.

Dismantling and safely securing nuclear and chemical weapons is a very expensive undertaking. Accordingly, in the quest to make the world safer and more peaceful, many industrial nations are heavily subsidizing Russia's dismantling efforts. France, Germany, Italy, Japan, and the United Kingdom have offered some $200 million in help related to nuclear disarmament over the next several years, and the United States has pledged even more. Under the Nunn-Lugar program, the Pentagon makes $400 million a year available to help Russia, Belarus, Ukraine, and Kazakstan dismantle their nuclear and chemical weapons. The United States has also agreed to purchase five hundred tons of highly enriched uranium that will be removed from Russian weapons over the next twenty years. The total cost is estimated at $11.9 billion.[6]

While the major nuclear powers are whittling down their stockpiles of nuclear arms, South Africa's new democratic government has destroyed entirely the nation's previously secret store of nuclear arms.

Meanwhile, two retired U.S. generals who once helped preside over America's nuclear arsenal now outspokenly advocate complete nuclear disarmament. They are Gen. Lee Butler, former head of the U.S. Strategic Command, and Gen. Andrew Goodpaster, former supreme allied commander in Europe, and they have been joined in their plea by sixty other former generals and admirals from nations around the world.

Respect for Humanity

You don't have to look back very far in history to realize that human beings have come a long way on the evolutionary trail. As the late Dr. Martin Luther King, Jr., observed:

> Man was born into barbarism when killing his fellow man was a normal condition of existence. He became endowed with a conscience. And he has now reached the day when violence toward another human being must become as abhorrent as eating another's flesh.

Consider a few historical facts that today seem almost beyond belief. Until the second half of the nineteenth century, human beings were almost routinely bought and sold as slaves. During the Industrial Revolution, small children in England and America toiled long hours in sweatshops and coal mines. Early in this century, women weren't allowed to vote in America. In 1954, the "separate but equal" doctrine

6. Ibid.

was overthrown in the United States. As a result, schools, restaurants, playgrounds, and other facilities open to the public became racially integrated. The United States has subsequently enacted laws ensuring equal opportunity for all.

Let's look at the strides women have made in American society since they gained the right to vote in 1920. Since then, women have been elected to the House of Representatives, the Senate, and governors' mansions in various states, as well as countless other governmental offices. Currently, more than 100,000 locally elected government officials are female, which amounts to over 20 percent of the total.

Women have made great strides as well in business. Nearly 6.4 million, or about one-third, of all U.S. businesses are owned by women.

The groundswell of opposition to "man's inhumanity to man" can be seen in the number of international human-rights covenants and conventions that have been agreed to in the last thirty years. All have been signed by a majority of nations.

Signatories to International Human Rights Pacts

Covenant or Convention	Number of signatories (out of 191 countries)
Economic, Social, & Cultural Rights (1966)	137
Civil and Political Rights (1966)	136
Elimination of Racial Discrimination (1969)	151
Elimination of Discrimination Against Women (1979)	154
Against Torture and Other Cruel, Inhuman, or Degrading Treatment or Punishment (1984)	106
Rights of the Child (1989)	189

SOURCE: *Human Development Report, 1996*

That nations increasingly are establishing higher standards of peaceful, charitable conduct reflects, of course, greater individual charity and love throughout the world.

As individual acts of kindness and charity have multiplied in the United States, there has been a corresponding drop in criminal activity. Nowhere has this decline been more dramatic than in New York City. For decades crime in America's largest city was on the increase, to the point where people were afraid to visit, much less reside there. But a new era has clearly dawned in New York. In 1996, *for the first time since 1968*, there were less than 1,000 murders. Moreover, the city's overall crime rate for major felonies—murder, rape, robbery, assault, burglary, grand larceny, and auto theft—plunged 16 percent, its third straight yearly decline. This brought the decline in major felonies since 1993 to 39 percent, a remarkable drop!

Still more encouraging is the fact that crime rates are falling in most other major U.S. cities, as well as in the nation as a whole. Understandably, political leaders are quick to claim the lion's share of credit for such declines, citing such things as improved policing. But there seems little doubt that the trend also reflects the same fundamental improvement in individual relationships that marks our more charitable and loving attitudes toward one another.

In sum, we should all be overwhelmingly grateful to live in this most exciting, prosperous, and peaceful time ever in world history.

14.

The Spirit

"Something is happening in America that as a journalist I can't ignore. Religion is breaking out everywhere."
—Bill Moyers in *USA Weekend*, Oct. 11–13, 1996

"For what shall it profit a man, if he shall gain the whole world, and lose his own soul?"
—Mark 8:36

- Religious organizations are the largest recipients of charitable giving in the U.S., receiving $60 billion in 1995.
- In 1990 overall attendance at U.S. sporting events totaled 388 million, while overall attendance at religious services exceeded 5 billion.
- Religious worship is increasing around the world.
- To attract members, American religious institutions are increasingly incorporating nontraditional elements into their worship services and their overall offerings—everything from jazz and rock music, dancers and drama, to health clubs, athletic complexes, publishing programs, singles nights, and day-care centers.

The Spread of Faith

Of all the encouraging trends that mark the closing years of the twentieth century, none is more heartening or more important than the remarkable spread of spiritual values. The resurgence of open religious worship in the once-Communist nations of Eastern Europe and in the former Soviet Union is well known. Less widely recognized is the recent growth of worship in nations like the United States with long traditions of religious freedom.

Consider a few statistics. In just the short period from November 1992 to December 1995, according to polling conducted by the Gallup Organization, the share of American families who say that religion is playing a more important role in their lives climbed to 38 percent from 27 percent. That works out to a remarkably sharp increase: some 40 percent in just over three years. Over a much longer span, Gallup reports a gradual rise in the percentage of Americans regularly attending church or synagogue services. The share attending such a service during the week prior to the poll stood at 42 percent in 1995, up from 37 percent in 1940 and 40 percent in 1972.

For perspective, it may be useful to compare attendance at religious services with attendance at sporting events. In 1990, overall attendance at U.S. sporting events totaled 388 million, while the overall attendance at religious services exceeded 5 billion. The membership rolls of nearly all the main religions have expanded in recent years. By one estimate, nearly 60 percent of the American people belong to some religious organization. Membership is especially strong in the South, particularly in Baptist areas; the Southwest, where Baptists and Roman Catholics are prevalent; the West-Central region, with Roman Catholics and Lutherans predominant; and Utah, home base for the Mormons.[1]

The table at right looks at how religious worship has recently grown around the world. It shows how the number of adherents to the major religions has risen briskly in just three years, from mid–1992 to mid–1995, the latest period for which information is available. It also shows how atheism has declined in the same period.

1. "Where They're Filling the Pews," *U.S. News & World Report*, July 15–22, 1996.

Growth of Believers
(in thousands)

	1992	1995
Christians	1,833,022	1,927,953
Moslems	971,328	1,099,634
Hindus	732,812	780,547
Buddhists	314,939	323,894
Atheists	240,310	219,925

SOURCE: *World Almanac and Book of Facts, 1997*

As encouraging as such numbers are, it is even more heartening to note the greater seriousness of today's believers. People nowadays tend to go to religious services for reasons that are more profound than was the case years ago when, as Peter Drucker has observed, attendance was often "steered by heritage, habit, and social status." Now, in contrast, attendance is usually "an act of commitment, and therefore meaningful. It is no longer an act of conformity. . . ."[2]

Religious belief is surging even in countries, such as China, where the political leadership opposes it. A "tidal wave of religious fervor is sweeping over China," according to one recent report. "Chinese are returning to religion with a vengeance. In a new age of economic prosperity and weak central government, the once sacrosanct Marxist-Leninist ideology has been eclipsed by a sweeping revival of spirituality."[3]

When the Communists gained power in China in 1949, an estimated 4 million Chinese considered themselves to be

2. Charles Trueheart, "The Next Church," *Atlantic Monthly*, August 1996.
3. Matt Forney, "Religion in China," *Far Eastern Economic Review*, June 6, 1996.

Christians. Now, by comparison, estimates show that some 90 million—more than twenty times the 1949 total—are Christians, though only about 15 million are registered as such on China's official church ledgers.[4] Other religions are flourishing as well. In the northwest province of Xinjiang, Moslems make up more than half of the region's population of some 16 million. Peasants in Hebei recently rebuilt a Taoist temple destroyed by Red Guards in the Cultural Revolution. Subsequently, thousands of worshippers followed the tradition of piling up new clothing there, creating within one day a mountain of three thousand garments.

Remarkably, this religious surge is occurring despite considerable opposition by Chinese authorities who, in 1994, laid down rigid new rules restricting missionary work and requiring places of worship to register. Violators are subject to prosecution, and people deemed to be members of "superstitious sects or secret societies that carry out counter-revolutionary activities" may be executed. In a recent four-month period, police in the coastal province of Zhejiang alone destroyed some fifteen thousand "unregistered" churches and temples.

Meanwhile, an extensive underground network of Catholics, Protestants, and other religious faiths is claiming hundreds of thousands of new worshippers in China every year. The Catholic network appears the most organized, with some sixty bishops pledging obedience to the pope and Vatican—though government strictures mandate that religious worship not be subject to domination by any individual or institution situated outside China's borders.

There are increasing signs that the ground swell of religion in China may be more than the authorities can easily handle. Recently, for example, some ten thousand Catholics, including two hundred priests and nuns, three bishops and two brass bands, conducted a pilgrimage to a village in Hebei

4. Ibid.

province. Police set up roadblocks to discourage the event, but the marchers managed to get around them. Later, the police stood on nearby buildings videotaping the proceeding, but made no effort to stop it.

Religious Institutions Reach Out

The use of marching bands during the pilgrimage in Hebei is but one example of a much broader global pattern. Religious institutions around the world are spurring the spiritual resurgence by appealing to worshippers in a variety of new ways. In the United States, among other places, churches are reaching out to younger people, for example, by playing rock music during services. Such is the demand that churches now constitute the fastest-growing market of Washburn International, an Illinois maker of rock-music sound systems. The $40 million company entered the church market only in 1994, but demand from churches already approaches 25 percent of its overall sales.[5]

Services at Willow Creek Community Church in South Barrington, Ill., feature amplified popular music and plays performed by members of the church's large staff. As many as twenty-five thousand people attend each week and the number keeps rising. In Hoboken, N.J., a twelve-member musical combo plays at masses held at Saints Peter & Paul Catholic Church.

The role of music in rising church membership can hardly be understated. Jack Heacock, a Methodist pastor, recalls arriving at First United Methodist Church in Austin, Texas, in 1973 and finding only an organist who played baroque music at the church's two services. "I suggested that we broaden the range a little, to hit a few more musical styles." he recounted recently. "We grew those two services

5. Susan Jackson, "Onward, Rock 'n Rollers," *Business Week*, April 8, 1996.

from under 500 people in 1973 to nearly 900 people in 1988," shortly before he left to work elsewhere.[6]

Many churches are also turning to market research techniques to build up their memberships. In 1969, the membership of Church on Brady, a Baptist church in East Los Angeles, had dwindled to only forty-five, as the neighborhood become increasingly Mexican. To reverse the decline, the church gathered demographic information about the neighborhood that was then used to target potential members with mailings and direct contacts. As a result, the average attendance at a service is about seven hundred, including many Mexicans.

Religious institutions are also striving to make it easier for their members to attend services. With the ever-increasing family use of autos to get around, for example, many churches are expanding their parking facilities.

Besides parking facilities, the Second Baptist Church of Houston, known locally as "Exciting Second," provides a wide range of facilities, serviced by five hundred staffers. On its forty-two-acre site, the church also provides a health club, a football field, and a day-care center for the use of its twenty thousand members. In addition, the church offers education courses, a special club for its seven thousand single members, and a full schedule of entertainment arranged after consultation with officials at Walt Disney World.[7]

A Jesuit priest and professor of theology, I should add, was recently appointed to the board of directors of Walt Disney Co., Disney World's parent. He is Leo J. O'Donovan, president of Georgetown University, who recently noted that there is "a recurrence of religious themes in popular entertainment because the American people remain an intensely religious people."[8]

---

6. Marc Spiegler, "Scouting for Souls," *American Demographics*, March 1, 1996.
7. Ibid.
8. "Priest on Disney Board Hopes to Bridge Gap," *The Record*, July 8, 1996.

As the use of Disney consultants indicates, many religious institutions are adopting a less formal atmosphere to gain new members. In Canada, the Centre Street Evangelical Missionary Church of Calgary encourages members to wear T-shirts and jeans at services. The church also recently launched a "Saturday Night Alive" service aimed at singles, including many single mothers. The average age of the worshippers is about thirty years.

At what point, you may ask, does the new informality of a religious service begin to detract from the sense of worship that is, after all, at the heart of the service? Murray J. Berenson, a physician who regularly attends the Middle Collegiate Church on Manhattan's Lower East Side, has seen its service change from quite formal gatherings to exceedingly informal affairs at which jazz musicians and dancers perform. At first, he questioned whether his church was moving in the right direction. "But then," he recently recalled, "I thought, 'What did the missionaries do?' They got people into hospitals, cured their broken bones or wounds, and then sold them the faith. We're just using music instead of medicine." Attendance on some Sundays, he adds, now approaches four hundred, up from only a couple of dozen people before the jazz and the dancers.[9]

The appeal of a less formal approach is evident in the remarkable growth across the United States of so-called megachurches, or churches with many thousands of members and many millions of dollars pouring in on their collection plates. One of the largest and fastest-growing megachurches is South Barrington's Willow Creek Community Church, which encourages casual dress to go with the pop music and plays. The New Life Community Church in Rathdrum, Idaho, now counts about 1,800 worshippers at its weekend services. As recently as 1990, its membership comprised only a handful of

9. Frank Bruni, "An Old-Time Religion Gets Some New Twists," *New York Times*, November 8, 1996.

couples meeting for Bible study in a private home. At the Life
Center Foursquare Church in Spokane, Wash., growth has
been almost as rapid. Its membership recently reached 1,800,
up from only 300 in 1990.

To accommodate its rapidly growing membership, the
Saddleback Valley Community Church in Mission Viejo,
Calif., is building a ten thousand-seat auditorium, a day-care
center, a fellowship hall, and an office headquarters on the
seventy-nine acres it owns. The eventual cost, estimated at
more than $50 million, will be paid by the church's approxi-
mately twelve thousand members.

As they gain more members, some larger megachurches
have even begun writing their own music and publishing
their own religious-education books. In addition, many have
started offering training sessions for pastors of smaller
churches nearby. In the process, networks of smaller churches
are forming around the largest megachurches. For example,
some one thousand churches pay $199 annually to be mem-
bers of the Willow Creek Church Association. "These
megachurches are becoming teaching churches, just as you
have teaching hospitals," remarks Leonard I. Sweet, chancel-
lor of United Theological Seminary in Dayton, Ohio.[10]

A recent change in the loan policy of the U.S. Small Business Administration has given a welcome boost to businesses
that encourage religious worship. In early 1996, the SBA
began providing low-cost loans to small businesses that
produce or sell religious goods. Previously, such firms were
prevented from obtaining such loans on the ground that the
government should not engage in such lending.

The advent of the electronic superhighway known as the
Internet is also helping religious institutions reach out to
new members through "pages" on the World Wide Web.

10. Gustav Niebuhr, "Power Shift Alters Way U.S. Protestants Worship," *New
York Times*, May 27, 1995.

Information available ranges from papal documents and sermons by well-known Protestant preachers to religious messages aimed at Buddhists, Moslems, Jews, Hindus, and Christians. On its Web page, for example, the Central Presbyterian Church of Montclair, N.J., refers to itself as the First Church of Cyberspace. Its page supplies news about the church's congregation and information about its library. The list of religious-oriented Web pages includes sites for the Christian Coalition, Billy Graham's archives, and such Roman Catholic orders as the Benedictines, Franciscans, Claretians, and Jesuits. The Internet also contains many writings by such early Christians as Bishop Augustine of Hippo, Ignatius Loyola, founder of the Jesuits, and Thomas à Kempis. Summing up the Internet's role, the Rev. John Rollins of Christ Episcopal Church in Pompton Lakes, N.J., recently noted that it provides yet another "means of communicating the good news of the Christian faith and provides links to the greater church throughout the world."[11]

Making Life Better

The spread of spirituality has brought other welcome developments that are working to make life better for people. Study after study has shown that as religion deepens within families, painful problems that once may have seemed intractable tend to ease or even disappear. Frequent U.S. churchgoers, for example, are about 50 percent less likely to suffer psychological problems and 71 percent less likely to be alcoholics than the general population, according to a recent survey. In addition, the divorce rate for weekly churchgoers is

11. Charles Austin, "Spreading the Word on the World Wide Web," *The Record*, July 7, 1995.

18 percent, far below the rate of 34 percent for those who attend a service less than once a year.[12]

"It's remarkable how much good empirical evidence there is that religious belief can make a positive difference" in the fight against drug abuse, alcoholism, crime and other social problems, remarks John Dilulio, a political scientist at the Brookings Institution, a nonprofit research organization based in Washington D.C. Churches, he adds, are "leveraging 10 times their own weight [in] solving social programs for us."[13]

Similarly, a recent study by Richard Freeman, an economist at Harvard, indicates that regular attendance at church is the best guarantee that a black youth in a U.S. city won't use drugs or turn to crime. It is a better guarantee, the study finds, than even income or family structure.

In its 1995 annual report, I should add, the Ford Foundation describes the church as "one of the African-American community's most important institutional resources." For this reason, the foundation funds a variety of church-based programs that deliver secular social services, and also backs efforts to convince other donors to do so.

Self-discipline, of course, is a central teaching of many religions, which helps explain such findings. For example, the five million or so U.S. followers of Islam, nearly half of whom are native-born blacks, are taught to abstain from alcohol and tobacco. One notable practitioner of such teachings is Rashaan Salaam, a Heisman trophy winner and Chicago Bears running back. His father, Teddy Washington, turned to Islam after a battle with alcohol and other drugs hastened his departure from the National Football League. The son, who prays five times a day, claims that, besides abstinence, "Islam

12. "Can Churches Save America?" *U.S. News & World Report*, September 9, 1996.

13. Ibid.

Extra-curricular activities, AAAS Black Church Project, funded by the Ford Foundation.

has taught me to be patient with people [and] to live life with a positive attitude."[14]

Inasmuch as spirituality discourages alcoholism and other drug abuse, it obviously fosters better health. But there also is growing evidence of linkages between spirituality and good health, quite apart from health benefits of the abstinence that some religions encourage. A recent survey finds, in this regard, that 79 percent of American adults believe that strong spiritual faith can help people recover from illness or injury. Moreover, 56 percent claim that their religious faith has in fact helped them recover from illness or injury.[15]

14. Andrew Hermann, "Islam a Growing Spiritual Force," *Chicago Sun-Times*, May 19, 1995.
15. Tom McNichol, "The New Faith in Medicine," *USA Weekend*, April 5–7, 1996.

Many doctors, to be sure, maintain that there is no scientific basis for a link between religious belief and health. However, recent studies show a growing amount of circumstantial evidence that deeply religious individuals by and large recover faster from illnesses and tend to be healthier. For instance, Dale Matthews, an associate professor of medicine at Georgetown University, has surveyed over two hundred cases to assess any ties between spirituality and health. In strongly religious patients, who made up about 75 percent of the sample, he detected distinct health benefits, such as a tendency toward lower blood pressure.[16]

Along the same line, a recent study at the University of Pittsburgh Medical Center found that heart-transplant patients with strong religious beliefs generally had less difficulty recovering from their surgeries and were healthier in the longer term as well. Interviews were conducted with 119 recipients of heart transplants, as well as their doctors and nurses. In general, the study concluded, the patients "who did have a stronger commitment to their faith and did use it more actively in their lives did better in the post-transplant years."[17]

As spirituality has spread in recent years, so has the volume of charitable giving, which many religions of course encourage. This generosity, in turn, serves to improve the lot of society's less fortunate members. Americans gave some $144 billion to various charities in 1995, for example, and much of this money—some $60 billion—was funneled through religious organizations to needy families. This was a record sum, up nearly 8 percent from the previous year. The increase, I should add, marked the sharpest yearly gain in giving in a decade. After religious organizations, the largest recipients of

16. Ibid.
17. Caryle Murphy, "Considering Religion's Role as an Influence in Recovery," *Washington Post News Service*, September 30, 1996.

charitable donations were educational groups (about $17 bil-
lion), health and human-service organizations ($12 billion
each), and the arts ($10 billion).

There is no way to pinpoint the precise degree to which
the spread of spirituality has contributed to a number of
heartening social trends in the United States. I strongly suspect, however, that it has been a factor in such developments
as these:

- In large U.S. cities, the number of violent crimes—murder,
  forcible rape, robbery, and aggravated assault—dropped 8
  percent in 1995, according to the Federal Bureau of Inves-
  tigation's latest annual crime survey, published in October
  1996. There were 13 percent less murders than in 1991
  and the murder rate was the lowest in a decade.
- The rate of births to teenage females in the United States
  declined in 1995 for the fourth straight year. The rate fell
  in forty-six of the fifty states. The birth rate for unmarried
  women also dropped, reversing a twenty-year pattern of
  increases.
- In 1994, the U.S. divorce rate stood at 4.6 per 1,000 population. This was the lowest level since 1973 and down from a
  peak of 5.3 in 1981.

A Mutually Beneficial Mix

A particularly heartening aspect of spirituality's rise in recent years is a growing sense that no conflict need exist between deeply held religious convictions and a steadfast faith
in the virtues of free-market capitalism, such as I maintain.
On the contrary, the two mix exceedingly well, I am convinced, tending in fact to supplement one another.

In the Introduction, I noted that dozens of books have appeared of late that emphasize the importance of ethics in

business affairs and explore linkages between religious teachings and the workings of the marketplace. They bear such titles as *The Seven Spiritual Laws of Success,* which was recently the best-selling business book in the United States, *The Management Methods of Jesus,* and *Jesus C.E.O.* A common thread is that no excuses for the profit motive are made in such books. Rather, they encourage entrepreneurship and discuss *modi operandi* that will help promote a company's good fortune, as well as improve its bottom line, without in any way compromising high ethical standards.

I offer my own experience as an investor as an indication of a clear linkage between religious belief and endeavor on the business front. If you make a basic effort to be in harmony with the Creator and all of His children through prayer, it is far likelier, I submit, that what you do in life, including your work effort, will turn out for the best. For more than forty years, all directors' meetings of Templeton mutual funds have begun with a prayer—not of course to pray that one or another investment would go up in price, but to pray that decisions made that day would prove wise and beneficial. I am convinced that more of our decisions at these meetings tended, as a result, to be sound than if there had been no prayer. And the reason, I believe, is that if you pray in this way, anything you choose subsequently to do will be done with a pretty clear mind. Your thinking won't be distorted by conflicts and you will be less likely to quarrel counterproductively with your associates.

True happiness, I should add, comes from spiritual rather than material wealth. Spiritual wealth leads to true happiness because with it we have a resource that will never run dry. Material wealth depends on factors often beyond our control, but spiritual wealth is entirely within our control. We alone determine how much or how little we have of it in our lives. If we take inventory and find ourselves lacking in spiritual wealth, it is up to us to draw on that great resource within us and replenish the supply. We have within us all that is necessary to make our lives satisfying, the ability to love and be loved.

This is not to say that material comfort cannot be a positive force in our lives. With it, we never need worry about going hungry or paying our bills or educating our children. But spiritual wealth is always there to serve us, through bad times as well as good times. It is our "blank check" that will be honored any time and any place. If we need wisdom, it is to that bank we go. If we lose all of our financial resources, our spiritual wealth will help us recover and recoup our losses. When spiritual wealth is the main security in our lives, we gain a deep, abiding peace that cannot be obtained with material wealth alone. In the process, we achieve true happiness.

That spirituality and business success are mutually beneficial is especially evident in the increasing numbers of successful business people who opt to alter their lives in order to make more room for spiritual pursuits. The average age of Protestant and Catholic theological students, for example, now exceeds thirty-four years, an increase of more than three years since 1991, and church officials attribute the increase largely to the rising share of students—65 percent, up from 61 percent—with previous careers in business or such professions as law.[18]

The trend, I should add, extends to many faiths. For instance, Ron is a Jew in his mid-forties with a doctorate in physics. He worked for the National Aeronautics and Space Administration and General Electric Co. before deciding recently to become a rabbi. He has a wife and five children, so the decision to become a rabbi, at least initially, has placed him under some financial pressure. However, he recently remarked that since the career change "I'm more in harmony with what I was intended to do." With his scientific

18. "More Are Trading Office for the Seminary," *The Record*, August 14, 1995.

background, he added, he has been working on a computer program to reconcile the Hebrew and secular calendars.[19]

The increasing blend of spirituality and business is also beginning to affect Hollywood, an industry not known traditionally as a source of religious messages. Indeed, Hollywood has often been criticized for focusing too much on sex and violence and too little on religious themes. But even in the movie business, with its understandable concern about producing films that will draw in audiences and make lots of money, I detect a growing interest in spiritual matters. In 1996, for instance, several movies were successfully released that offered uplifting, religious messages. They weren't the big-budget epics with big-name stars that Hollywood cranked out years ago in which oceans part and God's voice booms from on high. Rather, the new crop of religious films generally focus on everyday people with whom audiences can readily relate and even draw inspiration from.

An example is *The Spitfire Grill*. It tells the moving story of a young woman, an ex-convict, who struggles to start her life over again in a rural town and in the process brings a great deal of spiritual inspiration and joy to those around her. The film was produced for some $6 million by a Roman Catholic charity and was subsequently bought by Castle Rock Entertainment for $10 million after it had won an award at the Sundance Film Festival. Within several months of its introduction in theaters around the United States, the picture, clearly a financial success, had already earned more than $12 million.

The success of such a film is yet another heartening sign that spirituality constitutes a growing force in today's world, and that is yet another piece of good news for which we can all be very grateful.

19. Ibid.

15.

▬▬▬

The Future

▬▬▬

"There's a good time coming, boys, a good time coming."
—Charles Mackay (1814–1889), *The Good Time Coming*

No one knows what the future holds for humankind. Nor would I be so bold as to venture any flat-out predictions of what may lie in store for us. A long life convinces me, however, that to gain some sense of what may await us, we can profit greatly by paying close attention to past trends. I am not talking about headline-grabbing "trends" that span a few weeks or a few months or even a year or two. The sorts of trends I have in mind are much longer ones, trends that span decades, even generations, the sorts offered for your consideration in this book.

My lifelong interest in the longer term leaves me with a strong sense not only that the rate of progress is speeding up, but that this acceleration is apt to continue across countless areas of human endeavor in coming decades. Not every long-term trend, to be sure, is destined to persist indefinitely; were this the situation, we surely would be able to know the future. But most major trends do tend to persist, often gathering speed as they go along. And most, thank goodness, tend on balance to be beneficial. The proof, I submit, is in the remarkable strides recorded in this book.

So let me conclude this survey of human progress with a forward-looking summary of some of the glorious trends we have observed that seem likely to continue and possibly even accelerate in coming decades. It is true that we can't know the future, but we can say with assurance that if many of these trends do continue, there is indeed "a good time coming," as Charles Mackay aptly phrased it more than one hundred years ago.

The productivity of workers in the United States, as we have seen, has approximately doubled just since 1960, which only slightly exceeds the span of a single generation. In the same period, moreover, the real hourly pay of American workers has climbed roughly 60 percent.

Can we anticipate similar advances in the early decades of the twenty-first century, which is about to dawn? Why not? Indeed, why should we not look forward to even swifter gains in productivity and earning power? With the accelerating technological advances that modern science keeps providing, the prospect seems great indeed that worker productivity—and thus real pay levels—will keep on rising. Moreover, these gains most probably will quicken as technological strides that first spurred productivity on the farm and then on the factory floor increasingly benefit the global economy's fast-growing service sector.

Similarly bright is the outlook for housing. Just since 1980, the number of occupied housing units in the United States has expanded some 20 percent, so that there is roughly one home for every three people. This expansion constitutes an appreciably swifter increase than the rise of the U.S. population in the same period. In addition, as we saw in Chapter One, the median size of homes is on the rise. With the remarkable technological innovation that the home-building industry has achieved—from lighter, cheaper, sturdier materials to swifter, simpler construction methods—there seems every reason to expect these welcome gains in housing to continue. This appears likely not only in the United States but across vast regions of the world where for many people the dream of owning a home has only lately begun to materialize.

We traced in Chapter One the growing wealth of U.S. households. We saw not only that total wealth—tangible plus financial assets—has risen throughout the post-World War era, but that in the past couple of decades, the rise has been appreciably sharper than in the previous two. Just since the mid-1970s, in fact, wealth per U.S. household has approximately doubled! This was a time, you may recall, that witnessed repeated business recessions, corrosive rates of inflation, massive trade deficits, and Brobingnagian levels of red ink in the federal budget. Now, in a period distinguished by more persistent economic growth, mercifully milder inflation, more competitive businesses, and more nearly balanced budgets, is it overly optimistic to expect, at the least, another doubling of wealth during the next couple of decades? I think not. Indeed, an even sharper increase would hardly surprise me.

As wealth has increased, so has the nutritional well-being of most people. Thanks to technological advances and more sensible land management, fewer farms manage to provide ever greater quantities of food. The value of U.S. farm output, as we observed in Chapter Two, has expanded some six-fold just since 1950. With today's—and tomorrow's—agricultural advances, why should we not expect an output gain in the next half century that will match or, more likely, exceed the post-1950 increase?

Nor is this bright prospect confined only to the United States. As the world's developing nations leapfrog to the latest farming technologies, I suspect their levels of food production will climb even more sharply than gains recently achieved in the United States and other industrialized nations.

With better nutrition has come better health, and again there is every reason to expect the improvement to continue and very likely accelerate. Technology is advancing rapidly on many fronts, but nowhere have the strides been more remarkable than on the medical frontier. As we saw in Chapter Three, the five-year survival rate for U.S. cancer patients has climbed to more than 50 percent from only 20 percent in 1930. With our understanding of the disease expanding far more rapidly now than in earlier decades, it is reasonable to expect this encouraging increase in the survival rate to continue and even accelerate.

This encouraging trend in cancer survival, as well as the prospect of a further sharp improvement, is mirrored in countless other areas of health. The upshot, as we have noted, is that people are living longer, healthier lives. Some experts now foresee the average life expectancy of humans reaching 115 years sometime after the year 2050. That estimate, of course, is based on our present understanding of our bodies. But as medical research speeds forward in the years just ahead, perhaps even 115 years will prove an underestimation!

As the condition of our health continues to improve at a dramatic rate, so do the conditions under which we earn our livelihoods. We observed in Chapter Four how much safer and pleasanter the workplace has become. There seems every reason to expect that tomorrow's workplace will be even safer and more agreeable as automated work procedures continue to assume more and more of labor's most dangerous and tedious tasks. If present trends continue, as I suspect they will, more and more workers around the world will be free, thanks to labor technology, to pursue more challenging, interesting, well-paid jobs.

Advancing technology, quite obviously, is a crucial element in future achievement, whatever the field of endeavor may be. And, as we noted in Chapter Five, the pace of technological advance is speeding up remarkably. In the 1960s, the average lag was about thirty years between a basic discovery and its practical application, while now discoveries become obsolete in perhaps a few years, or occasionally even less.

As advancing science breaks through new barriers to a fuller understanding of our world, can there be any question that the technological pace will quicken even more? I noted in Chapter Five—to cite but a single remarkable instance of technological advance—that a U.S. car-rental firm already has installed in some eight thousand vehicles systems that display street maps, highlight routes to desired destinations, and offer voice instruction so drivers need not look away from the road. Only a few years ago, such automated navigators seemed unimaginable. But now, I find it difficult to imagine a car a dozen years hence without one!

Political and economic freedoms, of course, are intertwined. Both, as we have seen, have expanded substantially in recent decades. Is it fanciful to expect that the world's few remaining totalitarian regimes will move, as did the former Soviet bloc, toward greater political freedom? Recent history clearly points in that direction.

Nor is there likely to be a turning back in the global march to freer—and by no coincidence more prosperous—economics. The failure of central economic planning in the Soviet Union and elsewhere, as well as the success of such market-oriented nations as Chile, leaves little doubt that nations generally will continue to free their economies from excessive regulations, lofty tax rates, and other governmental burdens that tend to stifle economic growth and prosperity.

No field of human endeavor holds more promise, in my view, than education. The literacy rate in China, as I noted in Chapter Eight, has risen dramatically in recent decades—from less than 20 percent of the population in 1950 to more than 60 percent today. Can there be any question that this heartening rise will continue as strides in information technology bring the opportunity to learn to even the remotest parts of this vast Asian country? Is it overly optimistic to suppose that illiteracy will virtually disappear in China within another decade or two? I think not.

In the United States, though illiteracy is far less of a problem than in China or many other developing nations, information technology nonetheless plays a crucial role in the educational process. Some 80 percent of U.S. colleges, for example, now employ some combination of taped and live video instruction and fully half have two-way video link-ups. Such arrangements can only grow more common as information moves ever more easily and swiftly from place to place on our planet. The happy prospect is that more and more students— not only in the United States but around the world—will have access to the best possible standards of instruction.

The advance of information technology has indeed been startling, and the gains seem likely to continue and quicken. The power of the microchip, discussed in Chapter Nine, has recently been doubling approximately every eighteen months. At that rate, computers a dozen years from now will possess about 250 times the power of today's units! At the same time, by no coincidence, the amount of information available to us is doubling roughly every two and one half years—a rate that would bring a one thousand-fold increase within a quarter century!

These, to be sure, are staggering projections. But to any skeptic who questions whether these trends can continue, I point out that information has been proliferating at an accelerating rate for centuries. If a well-established past is any sort of prologue, the technological explosion in information and communications appears to have an almost boundless future.

Accelerating progress seems certain to continue as well in fields as diverse as transportation, the enjoyment of our leisure hours, and our efforts to safeguard and improve the natural environment around us. Consider, for example the progress made of late in America's effort to purify its rivers. As we saw in Chapter Twelve, some 60 percent of the nation's river-miles now meet federal standards for fishing and swimming, whereas as recently as 1970 the rate of acceptability was only 25 percent. As public concern over protecting the environment continues to grow, it seems reasonable to expect a still higher rate of acceptability. If the rate has risen from 25 percent to 60 percent in less than a generation, is it not possible that within another generation virtually all of the nation's many rivers will run cleanly?

Perhaps most heartening, as we consider the broad outlook for humankind, is the recent improvement in our ability to get along with one another and, not unrelated, our increasing attention to matters of the spirit. I need hardly observe that history in these areas of human experience seems a good deal less encouraging than it does in most other areas. The record is marked by repeated wars, cold ones as well as hot ones, often launched by godless, dictatorial regimes. However, there can be no denying that recently, as we observed in Chapters Thirteen and Fourteen, things have begun to change. Along with the spread of political and economic freedom, another most welcome trend is clearly taking hold—the spread of goodwill among people and between nations. Of all the heartening trends that we have examined in this book it is this one, I submit, for which we should be most grateful. For it is the one that makes the bright future that we have envisioned increasingly probable.

Bibliography

Books

Baumol, William J. and Alan S. Blinder. *Economics: Principles and Policy,* 5th ed. New York: Harcourt Brace Javonovich, 1991.

Douglas, Roger. *Unfinished Business.* Auckland, New Zealand: Random House, Ltd., 1993.

Easterbrook, Gregg. *A Moment On The Earth.* New York: Viking Books, 1996.

Gates, Bill. *The Road Ahead.* New York: Penguin, 1995.

Hastings, Paul. *Railroads: An International History.* New York: Praeger Publishers, 1972.

Human Development Report 1995. New York: Oxford University Press, 1995.

Malabre, Alfred L., Jr. *Beyond Our Means.* New York: Random House, 1987.

Matthews, Peter, ed. *The Guinness Book of Records 1996.* New York: Bantam Books, 1996.

McRae, Hamish. *The World in 2020: Power, Culture and Prosperity.* Boston: Harvard Business School Press, 1994.

McWhirter, Norris and Ross McWhirter, eds. *The Guinness Book of Records 1976.* New York: Bantam Books, 1976.

Moynihan, Michael. *The Coming American Renaissance.* New York: Simon & Schuster, 1996.

Naisbitt, John. *Megatrends Asia.* New York: Simon & Schuster, 1996.

Naisbitt, John. *Megatrends.* New York: Warner Books, 1984.

Negroponte, Nicholas. *Being Digital.* New York: Vintage Books, 1996.

Panati, Charles. *Panati's Extraordinary Endings of Practically Everything and Everybody.* New York: HarperCollins, 1989.

Schor, Juliet B. *The Overworked American: The Unexpected Decline of Leisure.* New York: Basic Books, New York, 1992.

Simon, Julian L. *The State of Humanity.* Cambridge, Mass.: Blackwell Publishers Inc., 1995.

The State of Canada's Environment, 1991. Ottawa: Ministry of Supply and Services, 1991.

Statistical Abstract of the United States.

World Almanac 1997. Mahwah, N.J.: World Almanac Books, 1997.

Wriston, Walter B. *The Twilight of Sovereignty.* New York: Charles Scribner's Sons, 1991.

Magazines

"A Computerized Overseer for the Truck Drivin' Man." *Business Week,* September 19, 1994.

Armstrong, Larry with Dori Jones Yang, Alice Cuneo, and *Business Week* bureau reports. "The Learning Revolution." *Business Week,* February 28, 1994.

Bierck, Richard. "Steering Clear of Danger." *U.S. News & World Report,* August 26, 1996.

"Can Churches Save America?" *U.S. News & World Report,* September 9, 1996.

Carr, Thomas, Heather Pedersen, and Sunder Ramaswamy. "Rain Forest Entrepreneurs." *Environment,* September 1993.

Dreifus, Claudia. "Peace Prevails." *New York Times Magazine,* September 29, 1996.

"Economic Freedom of the World, 1975–1995." The Fraser Institute, 1996.

Epstein, Gene. "The Demise of Job Security in the U.S. Is More Fiction Than Fact." *Barron's,* April 10, 1995.

"Financing World Growth." *Business Week,* October 3, 1994.

Forney, Matt. "Religion in China." *Far Eastern Economic Review,* June 6, 1996.

Hale, David. "Rethinking the World." *Barron's,* August 22, 1994.

"Hitchhiker's Guide to Cybernomics." *Economist,* September 28, 1996.

Holdgate, Martin W. "The Environment of Tomorrow." *Environment,* July–August, 1991.

Jackson, Susan. "Onward, Rock 'n Rollers." *Business Week,* April 8, 1996.

Jaroff, Leon. "Keys to the Kingdom," *Time* Special Issue, "The Frontiers of Medicine," Fall 1996.

Karatnycky, Adrian. *Freedom Review,* January–February 1996.

Lanser, Thomas R. "The Press in South Africa." *Freedom Review,* March–April 1996.

Lewis, Robert. "A Surprising Stability." *AARP Magazine,* April 1996.

Mahar, Maggie. "Russia's New Face." *Barron's,* June 10, 1996.

Mitchell, Russell and Otis Port. "Fantastic Journeys in Virtual Labs." *Business Week,* September 19, 1994.

Morin, Richard. "The Blame Game." *Washington Post Weekly Edition,* September 16–22, 1996.

Nakicenovic, Nebojsa. "Freeing Energy From Carbon." *Daedalus,* Summer 1996.

Norman, Michael. "Living Too Long." *New York Times Magazine,* January 14, 1996.

Nye, Joseph and William Owens. "America's Information Edge." *Foreign Affairs,* March–April 1996.

Peppers, Don and Martha Rogers. "As Products Get Smarter." *Forbes ASAP,* February 26, 1996.

Sands, David R. "No Whining Paradise." *Insight,* April 1–8, 1996.

Schrof, Joannie M. "The Winning Edge." *U.S. News & World Report,* July 29, 1996.

Smolowe, Jill. "Older, Longer," *Time* Special Issue, "The Frontiers of Medicine," Fall 1996.

Spiegler, Marc. "Scouting for Souls." *American Demographics,* March 1, 1996.

Stern, Caryl. "Who is Old?" *Parade Magazine,* January 21, 1996.

Sutell, Scott. "Companies Catching on to Telecommuting." *Crain's Cleveland Business,* October 30, 1995.

Tierney, John. "The Optimists Are Right." *New York Times Magazine,* September 29, 1996.

Trueheart, Charles. "The Next Church." *Atlantic Monthly,* August 1996.

Ubell, Earl. "What Medicine Will Conquer Next." *Parade Magazine,* November 5, 1995.

Wallechinsky, David. "Bad Predictions." *Parade,* September 10, 1995.

Wallechinsky, David. "Be at Home on the Internet." *Parade Magazine,* November 19, 1995.

"Where They're Filling the Pews." *U.S. News & World Report,* July 15–22, 1996.

Witkin, Gordon. "Can 'Smart' Guns Save Many Lives?" *U.S. News & World Report,* December 2, 1996.
Zellner, Wendy. "Team Player." *Business Week*, October 17, 1994.

Reports

Bank for International Settlements, 1994 Annual Report.
Calculations by Federal Reserve Bank of St. Louis. Reported in the *Wall Street Journal*, June 14, 1995.
Camdessus, Michel. "Social Dimensions of Economic Restructuring." *IMF Survey*, December 14, 1992.
Conservative Party Position Paper. January 19, 1996.
Department of Labor. Bureau of Labor Statistics. *Worker Displacement During the Mid-1990s.* USDL report 96-336, August 22, 1996.
Ratliff, William and Roger Fountaine. "Argentina's Capitalist Revolution Revisited." Hoover Institution, 1993.
Renner, Michael. "Budgeting for Disarmament." *Worldwatch Paper 122.*
The Rector's Report and *The Record.* St. Paul's School, 1994–1995.
Study by the Federal Reserve and Joint Economic Committee. Reported in the *Wall Street Journal*, July 11, 1995.
United Nations Development Programme's *Human Development Reports (1994–96),* Oxford University Press, New York.

Acknowledgments

Introduction

Blondie cartoon copyright © 1996. Reprinted with special permission of King Features Syndicate.

Chapter 2

Marie Antoinette woodblock courtesy of Corbis-Bettmann.
Malthus portrait courtesy of Corbis-Bettmann.
The Gleaners by Jean-Francois Millet courtesy of Corbis-Bettmann.
Photo of the world's largest pumpkin courtesy of AP/Wide World Photos.
"Rice and Wheat," *Catalysts for Cooperation,* United Nations Development Programme.

Chapter 3

World population chart from *The State of Humanity* by Julian L. Simon courtesy of Blackwell Publishers, Oxford, England.
Jenner statue photo courtesy of Corbis-Bettmann.
Photo of infant being vaccinated courtesy of Reuters/Corbis-Bettmann.
Photo of Dr. Barnard courtesy of UPI/Corbis-Bettmann.

Chapter 4

"Holding Ground" reprinted by permission of *Barron's*, © 1995, Dow
    Jones & Company, Inc. All rights reserved worldwide.

Chapter 5

Pickles cartoon © 1995 *Washington Post* Writers Group. Reprinted
    with permission.
"Molecular Structures" reprinted by permission of *Wall Street Journal*, © 1996, Dow Jones & Company, Inc. All rights reserved
    worldwide.
Apollo 11 photo courtesy of UPI/Corbis-Bettmann.
Photo of solar power plant courtesy of Kim Kulish/NYT Permissions.
"Revolutionizing Bridge Technology" reprinted by permission of
    Wall Street Journal, © 1996, Dow Jones & Company, Inc. All
    rights reserved worldwide.
Photo of Petronas Twin Towers courtesy of AP/Wide World Photos.
"The Superbike" courtesy of Rod Little, *U. S. News & World Report*.

Chapter 6

"The Map of Freedom" reprinted courtesy of Freedom House.

Chapter 8

Photo of QUASAR student reprinted by permission of Tony Gonzalez,
    © 1995.
Close to Home © John McPherson. Dist. of Universal Press Syndicate. Reprinted with permission. All rights reserved.
Photo of John Cleese courtesy of AP/Wide World Photos.

Chapter 9

Pepper . . . and Salt cartoon from the *Wall Street Journal* – Permission, Cartoon Features Syndicate.

"More Publishers Are Do-It-Yourselfers" courtesy of *New York Times* Graphics.

"Nations with Most Internet Computers" reprinted with permission of Knight-Ridder/Tribune Information Services.

New York Public Library photo courtesy of Ruby Washington/NYT Permissions.

"Irresistible Appeal?" reprinted by permission of *Wall Street Journal,* © 1996, Dow Jones & Company, Inc. All rights reserved worldwide.

Chapter 11

Barnes & Noble photo courtesy of Barnes & Noble and photographer Whitney Cox.

Bungee jumper photo courtesy of Reuters/Corbis-Bettmann.

Chinese McDonald's photo courtesy of Reuters/Corbis-Bettmann.

Chapter 12

"Freeing Energy from Carbon" figures reprinted by permission of *Daedalus,* Journal of the American Academy of Arts and Sciences, from the issue entitled "The Liberation of the Environment," Summer 1996, Vol. 125, No. 3.

Chapter 14

Photo courtesy of the AAAS Black Church Project, funded by the Ford Foundation.

Index

A

Academic Systems, Inc., 157
Accor SA, 208
adaptive optics, 93
advertising
　on Internet, 181–183
　on television, 176
Africa
　capitalism in, 132
　democracy in, 112
　democratic supports in, 122–123
　health in, 55–57
　literacy rates in, 161
African Americans, and religious obser-
　vance, 252, 253*f*
age, and capability, 50
agriculture
　future of, 261
　genetic research and, 39–42
　history of, 31–32
　injuries in, 70
　productivity and, 20, 34–35, 165
　progress in, 32–34
　technology and, 32, 38–41
　workers in, 32, 34, 34*t*
AIDS, drug therapy for, 63
airlines
　complaints against, trends in, 189
　employee involvement in, 79
airmail, 169
air quality, 213–215, 214*f*
Air Quality Accord (United States-
　Canada), 227
air transport, progress of, 187–189, 189*t*
Akashi Kaikyo Bridge (Japan), 96

Albendazole, 54
Aldrin, Edwin M., 92*f*
Algeria, democracy in, 120
Alzheimer's disease, 66
Amazon basin, forests of, 218
Amazon.Com Inc., 183
Amdahl Corp., 149–151
American Council on Education, 156–157
American Cyanamid Co., 56
American Management Association,
　165–166
American Stock Exchange, 142
Anderson, W. French, 64
Andy Warhol Museum, 206
angioplasty, 59
Angola, peace in, 230
Annenberg, Walter, 151
annuity payments, 145
answering machines, 174
apartheid, 126
Apollo missions, 91–92
Aquino, Corazon, 119
Arab countries
　democracy in, 120
　literacy rates in, 161
architecture, 97–99
Argentina
　free trade in, 140
　private investment in, 139
　privatization in, 135
Armstrong, Neil, 92
Asia
　forests of, 218
　leisure time in, 207
　tourism in, 200

Asian Development Bank, 137
Associated Press, 169
ATMs (automated teller machines), 101
Australia, and Internet, 181*f*
Austria, parental leave in, 195
automated teller machines (ATMs), 101
Automobile Manufacturing Association,
    186
automobiles, 186–187
    and air quality, 213
    with computerized navigation system,
        104–105, 190
    registrations in United States, 188*t*
    religious institutions and, 248
    and tourism, 201
automobile safety
    technology and, 106–107
    trends in, 186–187
Avery, Dennis, 32
AZT, 63

B
baby boomers, future of, 29
Bali, tourism in, 200
Ballistic Missile Defense Organization,
    93
banking
    in New Zealand, 118
    technology and, 101
bar code scanners, 103
Barnard, Christiaan, 59, 60*f*
Barnes Foundation, 206
Barnes & Noble, 198, 199*f*
Baumol, William, 211, 213
Baylor Research Institute, 93
Belarus, disarmament in, 234–235
Belgium
    labor costs per output in, 77
    parental leave in, 195
Bell, Alexander Graham, 173
Bell Atlantic Corp., 158, 175–176
Berenson, Murray J., 249
Berry, Jerry, 183
beta-blocking drugs, 59
Bettman Archive, 206
Bible, 229, 243
Biogen, Inc., 66
Blaiberg, Philip, 60*f*
Blomquist, William, 159

Boeing merger, 233
Bolivia, privatization in, 134
book publishing
    inspirational, 10, 255–256
    personal computers and, 178, 179*t*
    pessimism in, 12–13
book stores, 198, 199*f*
Boston Harbor, water quality in,
    215–216
Boston Stock Exchange, 142
Bourlag, Norman, 39
brachytherapy, 62
Brandeis University, 24
Brazil
    conveniences in, 28
    free trade in, 140
    physician availability in, 27
    private investment in, 139
    privatization in, 135
    telephones in, 175*t*
breast cancer, genetic research and, 65
Brecht, Bertolt, 31
bridge construction, progress in, 96, 97*f*
Britain. *See* United Kingdom
Broderbund Software, 156
Brody, Kenneth D., 133–134
Bronze, Bobby, 203*f*
Brookings Institution, 252
brown plant hopper, 42
Brown University, 18–19
bungee jumping, 202, 203*f*
Bureau of Labor Statistics, 7
Burma, future of, 115
Bush, George, 239
business. *See also* private business;
        small business
    education in, 153
    religion and, 255–258
    technology and, 99–100
    women in, 241
Business Information Systems, 171
Business Week, 89
Butler, Lee, 240

C
cable television, 176, 202
    in Asia, 210
California, air quality in, 214–215

call waiting, 174
calories consumed
  in emerging nations, 35
  meat/plant proportion and, 46–47
  in United States, 67
Cambodia
  peace in, 230
  peacetime budget of, 236–238
Camdessus, Michel, 140
camping, 201
Canada
  air quality in, 214
  college-entrance ratio in, 152
  and Internet, 181*f*
  as market for United States, 139
cancer, 58, 261–262
capital
  international flow of, 136–140
  per worker, 80–81, 81*f*
capitalism, 127–135
  religious observance and, 255–258
carbon, and energy consumption,
    222–225, 224*f*
Caribbean, enrollment in, 161
Caribbean Tourist Organization,
    200–201
Carnegie Mellon University, 107
Carson, Rachel, 212–213
Castro, Fidel, 115
CAT scans, 61
CD-ROMs
  educational, 155
  museums on, 206
cellular telephones, 174–175
centenarians, 49–50
Center for Global Food Issues, 32
Central African Republic, 57
central planning, 129–130
Central Presbyterian Church (Mont-
    clair, NJ), 251
Centre Street Evangelical Missionary
    Church (Calgary, Canada), 249
CFC (chlorofluorocarbon) emissions,
    214
Chad, democracy in, 120
Chamberlain, Louise, 158
Channel Tunnel, 96
charitable foundations, 9
charitable giving, 4, 254–255

charity
  and food, 47
  individual, 242
Chicago Stock Exchange, 142
child survival, advances in, 52–57
Chile
  child survival in, 53
  democracy in, 115–118
  economic freedom in, 132–133
  life expectancy in, 53
  privatization in, 135
  telephones in, 175*t*
China
  cellular telephones in, 175
  conveniences in, 28
  defense industry conversion in, 234
  education in, 153, 162
  energy consumption in, 95, 137
  food in, 27
  leisure time in, 207–208
  life expectancy in, 50
  literacy rate in, 147
  markets in, 137–138
  prosperity in, 26
  religious observance in, 245–247
  telephones in, 173
  televisions in, 28
  and tourism, 208
CHIPS (Clearing House International
    Payments System), 136
chlorofluorocarbon (CFC) emissions,
    214
cholesterol
  advances against, 61
  nutrition and, 67
Christ Episcopal Church (Pompton
    Lakes, NJ), 251
Christianity, in China, 245–247
Chrysler Building, 99
Churchill, Winston, 49
Church on Brady (Los Angeles, CA),
    248
Cincinnati Stock Exchange, 142
City University of New York, 151
class, wealth and, 24–25
Clean Water Act of 1972, 216
Clearing House International Payments
    System (CHIPS), 136
Cleese, John, 166*f*

Clinton, Bill, 24
administration, job growth during, 76
Colt Manufacturing Co., 102
Columbia, property rights in, 124
comfort, 21–25
communications, 167–184
advances in, 167–172
Communism, fall of, 113, 129, 229
information and, 231
compensation, 7
fringe benefits in, 84–85, 85t
future of, 260
competition, 76–77
Comprehensive Test Ban Treaty, 239
Computer Intelligence Infocorp., 180
Computer Intelligence Inforcorp., 183
computerized axial tomography, 61
computerized navigation system,
104–105
computer power
advances in, 177–179
future of, 264
Moore's Law of, 168
computers
advances in, 91
and business, 99–100
and education, 154–157
and entertainment, 205–207
personal. See personal computers
and photography, 105–106
and technological development, 89
conservation, 219–221
construction, 95–99
Consultative Group on International
Agricultural Research, 40
continuing education, 164–166
earnings and, 164, 164t
conveniences, 21–22
electrical, 93–94
in emerging nations, 28
prevalence of, 17–18
Convention on International Trade in
Endangered Species, 227
Cooper, Richard, 83
copying machines, 171
Corbis Corp., 206
coronary artery bypass surgery, 59
correspondence courses, 159

Costa Rica
peacetime budget of, 235
rain forest preservation in, 218–219
couch potatoes, 196
credit cards, 100
crime rates, decline in, 242, 255
Cuba, 113
future of, 115
peace in, 230
tourism in, 201
cultural associations, 207
Czech Republic
economic freedom in, 128
privatization in, 135
property rights in, 124–125

D
Daedalus, 222
Dalai Lama, 231
Damrongchaitam, Paiboon, 210
DDT, 213
deaths, in workplace, 69–70
debit cards, 100–101
defense industry, consolidations in,
232–235
defense spending, reduction in, 229, 232,
233t, 237t
demilitarization expenditures, 238t, 239
democracy, 8
advance of, 111–115, 120–124,
229–230
in Africa, 112
and partial freedom, 120–121
and peace, 120, 231
supports of, 120–121
dendrimers, 90
Denmark
and Internet, 181f
student-teacher ratios in, 152
dentistry, 63–64
desert locust, 43
desktop publishing, 178, 179t
digital photography, 106
digital science, 89
Dilulio, John, 252
Diners Club, 100
diphtheria, 54
disarmament, 232, 234–235

Disney World, 102
disposable income, 17
Disraeli, Benjamin, 193
distance learning, 158–161
divorce, trends in, 255
Doswell generating station, 225
downsizing
  in defense industry, 234
  as percentage of employment, 75–76
  perspectives on, 5–6, 74–76
Drucker, Peter, 245
drug therapy, 62
Drury College, 164
DuPont & Co., 56

E

earnings
  versus compensation, 7
  and continuing education, 164, 164t
  productivity and, 20
Earth Summit, 227
Easterbrook, Gregg, 225
Ebert, Roger, 91
economic freedom, 8, 127–145
  evaluation of, 128
  future of, 263
  increases in, 128–129
economic insecurity, 5
Ecuador, rain forest preservation in, 218
education, 4, 147–166
  in business, 153
  continuing, 164t, 164–166
  future of, 263–264
  increased enrollment in, 151–154
  Internet and, 157
  military cutbacks and spending on,
    235–239
  private aid to, 149–151
  progress in, 147–148
  spending on, 148, 149t
  technology and, 7, 154–161, 156t
  time spent on, 162
  work time and, 18–19
Egypt
  future of, 115
  telephones in, 175t
Ehrlich, Paul, 62

electricity, 93–95
  production of, 94
electronic funds transfer, 100
electronics, 101–104
Eli Lilly, 218
El Salvador, peace in, 230
e-mail, 170
emerging nations
  calories consumed in, 35
  conveniences in, 28
  energy consumption in, 221–225
  food in, 27
  increased enrollment in, 161–163
  investment in, 133
  land reform in, 45–46
  life expectancy in, 27, 49–50
  literacy rates in, 27
  physician availability in, 27
  prosperity in, 26
  unfilled needs in, 28–29
Emery, Ray, 88–89
Empire State Building, 99
employees. See workers
employee stock-ownership plans
  (ESOPs), 79–80
employment, after layoff, 76
Employment Policy Foundation, 80
endoscopy, 62
energy consumption
  in China, 137
  decarbonization of, 222–225, 224f
  global trends in, 95, 221–225, 223f
  in United States, 93–95
England. See United Kingdom
ENIAC, 178
enrollment
  increases in, 151–154, 161–163
  in secondary education, 148
entertainment, 202–205
  spending on, 202, 204t
environment, 211–227
  food production and, 46
  future of, 264–265
  getting along on, 225–227
  past problems in, 212
  perspectives on, 211–213
Epivir (3TC), 63
epoxy resins, 90

equal opportunity, 9, 85, 241
Eritrea, peace in, 230
Ervin, Robert, 190
Ethiopia, peace in, 230
ethnic cuisines, 37–38
European Union, 237
Evaluation Systems, 107
EZ Pass, 102–103

F
farms, changes in, 20, 34, 34*t*
fax machines, 171–172
Federal Highway Administration, 185
fertilizers, 42
   and water quality, 215
fiber optic lines, 172
films
   religious, 258
   spending on, 202
fingerprint scans, electronic, 102
Finland
   cellular telephones in, 175
   and Internet, 180, 181*f*
   leave policies in, 194
First Step program, 151
First United Methodist Church (Austin, TX), 247
fish farms, 43
Flavr Savr, 44
Fleming, Alexander, 62
fluoride, 63
food, 31–47
   distribution of, 47
   in emerging nations, 27
   in former Communist nations, 45
   future of, 45–47
   nutritive value of, 38, 41–45
   perspectives on, 3–4
   and population, 35
   prices of, 44–45
   sufficiency of, 35
   technology and, 6–7
   variety of, 35–38
Forbes magazine, 23
Ford Foundation, 147, 149, 252
foreign language study, 163
forests, 218–219
   in Massachusetts, 212
   productivity and, 20

former Communist nations
   capitalism in, 129
   democracy in, 113
   food in, 45
   and pollution, 221
   property rights in, 124–125
   religion in, 9–10, 243–244
Forrester Research, 184
Fosamax, 63
France
   food variety in, 37
   labor costs per output in, 77
   paid vacation time in, 194
   service jobs in, 78
franchise, 126, 241
Fraser Institute, 128–129
Freedom House, 112, 124, 127–128
Freeman, Richard, 252
free markets, 132–133
free press, and democracy, 121–123
Friedman, Milton, 111
fringe benefits
   growth of, 84–85, 85*t*
   and leisure, 194–195
   and real compensation, 7
Frito-Lay, 81–82
Frum, David, 24–25
fuels, carbon in, 224–225
Fujitsu, 151
future, 259–265
   of food, 45–47
   of life expectancy, 67, 262
   of living standards, 28–29
   of medicine, 64–67, 261–262
   of political freedom, 115, 263

G
gall bladder removal, 62
Gallup Organization, 244
Galton, Francis, 101
Gambia, child survival in, 53
gasoline, and air quality, 215
Gates, Bill, 106, 151, 179, 183–184, 206
GDP. *See* gross domestic product
Geingob, Hage, 162
General Services Administration, 83
genetic research, 90
   and agriculture, 39–42
   and cancer, 58

and environment, 109
future of, 46
in medicine, 64–66
on pest control, 43
on poultry, 43–44
on tomatoes, 44
Georgetown University, 248, 254
Germany
defense industry conversion in, 234
labor costs per output in, 77
parental leave in, 195
Gertel, Martin M., 84
getting along, 4–5, 9, 229–242
and environment, 225–227
future of, 265
Ghana, 112
glasnost, 131
The Gleaners (Millet), 38, 39*f*
Global Environment Facility, 227
Global Positioning System, 91
global trade, 136–140
and food variety, 35–37
impediments to, 140
goiter, 67
Goldsmith, Oliver, 147
Goodpaster, Andrew, 240
Gorbachev, Mikhail, 130–131
Gore, Al, 178
grades, improvement in, 153–154
Great Northern Paper Co., 71
green revolution, 39, 46
gross domestic product (GDP), rise in, 16–17
Guggenheim Museum, 207
Guinea worm, 56
Gurirab, Theo-Ben, 162
Gutenberg, Johann, 169

H
Hale, David, 136
happiness, 256–257
Harvard University, 151, 252
Hayek, Friedrich, 129–130
Heacock, Jack, 247–248
health, 4, 49–67. *See also* medicine
future of, 261–262
nutrition and, 66–67
religion and, 10, 253–254
spending on, 57, 235–239

Hearst, William Randolph, 169
heart transplants, 59
religion and, 254
Hertz Corp., 105
Hezel Associates, 158–159
high blood pressure, 59
higher education
in Britain, 148
computers and, 156–157
exchange students in, 163, 163*t*
expansion of, 153–154
increased enrollment in, 152, 161
military recruiting and, 165
highway system, 185
hobby clubs, 207
home learning, computerized, 155
home ownership, 21
home, technology and, 104
housing, 21
future of, 260
Hudson River, water quality in, 216–217
Hugo Boss, 151
Human Development Report, 161
Human Genome Project, 65
humanitarian aid, money for, 237–238
human rights, progress in, 240–242, 241*t*
hunger, 32. *See also* food
Hussein, Saddam, 115
hydroelectric power, 94

I
IBM, employee training in, 165
illness
advances against, 50, 51*t*
decline in, 49–52
religion and, 10
immunization, 52, 53*f*, 54, 55*f*
improvements in, 54
income
decline in, perspectives on, 17
disposable, 17
service jobs and, 77
India
conveniences in, 28
demographics in, 138
energy consumption in, 95, 223, 223*f*
food production in, 39–40
private investment in, 139

India *(continued)*
  prosperity in, 26
  telephones in, 175*t*
  tourism in, 200
Indiana-Purdue Universities, 159
individual enterprise, 8
individual rights, spread of, 124–126
Indonesia
  agriculture in, 42
  cellular telephones in, 175
  conveniences in, 28
  demographics in, 29, 138
  food in, 27
  future of, 115
  literacy rate in, 27, 148
  physician availability in, 27
  prosperity in, 26
  telephones in, 175*t*
  tourism in, 200
inequality, 25
inflation, in New Zealand, 119
information
  increases in, 168
  and peace, 231
information technology, 167–184
  advances in, 167–172
  future of, 263–264
inspirational books, 10, 255–256
insurance companies, as investment,
    144–145
Intel Corp., 168, 173, 220
intellectual property, 124–125
Intelligent Parking Meters, 101
interactive video disks, educational, 155
Inter-American Development Bank, 138
Intergovernmental Panel on Climate
    Change, 227
International Finance Corporation, 132
International Maize and Wheat Im-
    provement Centre, 39, 42
International Monetary Fund, 140
International News Service, 169
International Rice Research Institute,
    42
Internet, 87–88, 179–184
  access to, 158, 180–181, 182*f*
  and college search, 157–158
  communications function of, 173–174
  and education, 157

  museums on, 206
  and religious observance, 250–251
  as sales tool, 181–184, 184*t*
Inter-Parliamentary Union, 111
Inverni della Beffa, 218
investment
  in emerging nations, 133
  growth in, 142*t*
  with limited means, 9, 136–137,
    140–141, 143–145
  opportunities for, 140–145
  reduction of barriers to, 139
  and working conditions, 80–82, 81*f*
iodized salt, 67
Iraq, future of, 115
Isaiah, Book of, 229
Islam, 252–253
  in China, 246
Italy
  labor costs per output in, 77
  paid vacation time in, 194
i2 Technologies, 100
ITT Automotive, 177
Ivermectin, 56

J
Jamaica, privatization in, 134
Japan
  conveniences in, 28
  demographics in, 29, 138
  education in, 152–153
  investment in China, 138
  labor costs per output in, 77
  leisure time in, 207
  as market for United States, 139
  service jobs in, 78
  television in, 208
Jarvik, Robert, 59
Jenner, Edward, 52, 53*f*
job security, 73–78, 75*f*
Johnson, Samuel, 167, 184
journalists, abuse of, 123
judiciary, and democracy, 121, 123–124
juku system, 153

K
Kaczynski, Theodore J., 6
Karatnycky, Adrian, 112–113
keyboard, 169

Khan, Enkhasai, 113
Kilgallen, Jimmy, 169
Killen & Associates, 183–184
King, Martin Luther, Jr., 240
Klaus, Vaclav, 128
knowledge, increases in, 168

L
land, 218–219
    reserved for leisure use, 201–202
land reform, in emerging nations, 45–46
laparoscopy, 62
Latin America
    democracy in, 113
    enrollment in, 161
    free trade in, 140
    as market for United States, 138–139
Latvia, property rights in, 125
law enforcement, technology and, 102
leaf blight, 42
leisure, 8, 193–210
    global trends in, 207–210
    land reserved for, 201–202
    and life enrichment, 196–202
    perspectives on, 80–81
    technology and, 7
Lenfant, Claude, 59
libraries
    advances in, 158
    and Internet access, 180–181, 182*f*
Life Center Foursquare Church
        (Spokane, WA), 250
life enrichment
    leisure time and, 196–202
    religious observance and, 251–255
life expectancy, 49–67
    in emerging nations, 27, 49–50, 53
    future of, 67, 262
    and retirement, 195
    technology and, 7
Limited Test Ban Treaty, 212
Link Resources, 83
Lin, T.Y., 96
Liposorber, 61
literacy rates
    and agricultural productivity, 165
    in China, 147
    in emerging nations, 27
    future of, 263–264

    increases in, 161
    in Indonesia, 148
    in Thailand, 159
living standards, 15–29
    in Africa, 132
    and comfort, 21–25
    future of, 28–29
    global trends in, 25–28
    and peace, 231
    perspectives on, 15–16
    rise in, 15–19
London, pollution in, 4, 212
Lopez, Perry and Monica, 183
Lovell, James A., Jr., 91
Lozada, Gonzalo Sanchez de, 134
Lynch, Peter, 5–6

M
Mackay, Charles, 259–260
magnetic resonance imaging, 61
mail delivery, 169–170
Maine 200, 70–71
malaria, 56–57
Malaysia, tourism in, 208
Malthus, Thomas R., 35, 36*f*
Mandela, Nelson, 123, 126
Manley, Michael and Norman, 134
Mante, Tei, 132
manufacturing, reduction of pollution
        from, 219–221
Map of Freedom, 115, 116*f*–117*f*
March of Dimes, 52
Marie Antoinette, 33*f*
Mark, Book of, 243
market research, religious institutions
        and, 248
Mars, 91
Massachusetts Institute of Technology,
        156
    Media Lab, 177
Matthews, Dale, 254
Max Planck Institute of Biochemistry,
        66
McColl, Hugh L., 101
McCormick, Cyrus, 38
McDonald's, in China, 207–208, 209*f*
McDonnell Douglas merger, 233
McRae, Hamish, 46
measles, 54

media
  accessibility of, 11
  and hunger, 32
  pessimism in, 2, 11–12
  and spirituality, 258
medicine, 57–64. *See also* health
  future of, 64–67, 261–262
  progress in, 52
  research in, spending on, 57
  work time and, 18–19
Mediterranean fruit fly, 43
megachurches, 249–250
Merck & Co., 56, 218–219
Mercosur, 140
Messick, Richard E., 127
Mexico, unfilled needs in, 28, 138
microchip, future of, 264
Microsoft, 104, 106, 151, 179, 183, 206
Middle Collegiate Church (New York,
  NY), 249
Middle East
  democracy in, 120
  peace in, 230
military
  and democracy, 121
  personnel on active duty, 236*t*
military cutbacks, 229, 232, 233*t*, 237*t*
  and health and education spending,
  235–239, 237*t*
Millennium Broadway, 151
Miller, Herman, 100
Millet, Jean François, 38–39
Milton, John, 167
molecules, newly-invented, 90
money, technology and, 100–101
Mongolia, democracy in, 113–115, 120
Montreal Protocol, 227
Moore, Gordon, 168
Moore's Law, 168
Moyers, Bill, 243
Mozambique, peace in, 230
MRI (magnetic resonance imaging), 61
Museum of Modern Art, 207
Museum of Paleontology (Berkeley, CA),
  206
museums, 206–207
music
  religious institutions and, 247, 249
  youth population and, 205

mutual funds, 143–144
  growth in, 145*t*
  as percentage of individual financial
  holdings, 144*f*

N
Naisbitt, John, 168, 207
Nakicenovic, Nebojsa, 222
Namibia, 162
  peace in, 230
National Aeronautics and Space Admin-
  istration, 190
National Association of Securities Deal-
  ers, 142
  Automated Quotation System,
  142–143
National Basketball Association, 197
National Cable Television Association,
  180
National Educational Radio Network
  (Thailand), 159
National Football League, 197–198
National Hockey League, 198
National Institutes of Health, 65
National Park System, 201
National Sporting Goods Association,
  197
National Weather Service, 91
National Wildlife Refuge System, 202
NationsBank Corp., 101
natural gas generating station (Doswell,
  VA), 225
navigation system, computerized,
  104–105
Negroponte, Nicholas, 177
Nepal, education in, 162
Netherlands
  and Internet, 181*f*
  paid vacation time in, 194
Newark International Airport, 198–199
New Jersey Board of Education, 163
New Life Community Church (Rath-
  drum, ID), 249–250
Newman, Paul, 151
New York City, crime in, 242
New York Public Library, 180–181, 182*f*
New York Stock Exchange, 141–142
New York Times, 23, 216
New York University, 195

New Zealand
  economic freedom in, 132
  and Internet, 181*f*
  political freedom in, 118–119
Nicaragua, peace in, 230
Nkrumah, Kwame, 112
Norman, Michael, 195
North American Van Lines, 107
Northern Ireland, peace in, 230
Norway, and Internet, 180, 181*f*
Novak, Michael, 25, 111
nuclear power, 94
nuclear weapons
  control of, 212–213, 230
  disarmament and, 239–240
nurses, 57
nutrition, 38, 41–45. *See also* food
  future of, 46–47, 261
  and health, 66–67
Nye, Joseph, 87

O
Oates, Wallace, 211, 213
Ocamp, Roberto de, 119–120
Occupational Safety and Health Administration (OSHA), 70–71
O'Donovan, Leo J., 248
Oklahoma State University, 156
Olympic games, technology and,
  107–109
onchocerciasis. *See* river blindness
oncogenes, 58
optimism, 1–2
  effects of, 3
  media and, 12–13
  trends supporting, 5–10
organ transplants, 59
OSHA (Occupational Safety and Health
  Administration), 70–71
osteoporosis, 63
  genetic research and, 65–66
overwork, 80
Owens, William, 87

P
Pacific Mutual Life Insurance Co., 85
Pacific Stock Exchange, 142
pagers, 174

paid vacation, 194
Pakistan, educational spending in, 148
Paraguay, free trade in, 140
Paraziquantel, 56
parental leave, 194–195
Paris Peace Agreement, 236
Parkinson's disease, 66
Pathfinder, 91
PCBs (polychlorinated biphenyls), and
  water quality, 217
peace, 9, 229–230
  foundations of, 230–231
Pelisson, Gerald, 208
penicillin, 62
PepsiCo Inc., 81–82
perestroika, 131
Permethrin, 57
personal computers, 171
  and photography, 105–106
  and publishing, 178, 179*t*
  as television, 177
Peru
  privatization in, 135
  property rights in, 124
  rain forest preservation in, 218
pessimism
  effects of, 2–3
  on living standards, 17
  persistence of, 10–13
  perspectives on, 5
  public awareness of, 11–12
pesticides, 42, 213
  and water quality, 215
pests, advances against, 42–43
Petronas Towers (Kuala Lumpur), 98*f*,
  99
pets, radio-frequency identification of,
  103
pharmaceutical research, and rain forest preservation, 218–219
Philadelphia Stock Exchange, 142
Philippines
  free trade in, 140
  political freedom in, 119–120
Phillips, Barbara, 176
photography, 105–106
physician availability
  in emerging nations, 27
  in United States, 57

PictureTel, 177
Pinochet, Augusto, 115–118
Poland, food production in, 45
polio, 52, 54
political freedom, 8, 111–126
  approaches to, 115–120
  evaluation of, 114
  future of, 115, 263
  global trends in, 115, 116f–117f
  growth of, 111–115
politics
  and food, 45, 47
  television advertisement and, 176
pollution. *See also* environment
  perspectives on, 4
  prevention of, 219–221
  technology and, 7
  of water, 212
polychlorinated biphenyls (PCBs), and
    water quality, 217
polyethylenes, 90
Pony Express, 169
population, 51f
  food and, 35
  growth in, causes of, 50
Portugal, peacetime budget of, 235
poultry
  genetic research on, 43–44
  prices of, 44–45
poverty, perspectives on, 3
Precision Fabrics Group, 56
price controls, 129, 131
private business, 9
  and education, 165
  multinational, and peace, 231
  in United States, 141
private property
  and food production, 45–46
  and freedom, 124
privatization, 8, 129, 133–135
  in Columbia, 124
  and food, 45
  and telephone service, 173
produce, spoilage and, 44
productivity
  and agriculture, 20, 34–35, 165
  central planning and, 130
  employee involvement and, 79
  future of, 260
  and living standards, 19–20

  medicine and, 58
  technology and, 69, 72
progress
  in agriculture, 32–34
  in air transport, 187–189, 189t
  in bridge construction, 96, 97f
  in child survival, 52–57
  against cholesterol, 61
  in communications, 167–172
  in computer power, 91, 177–179
  in education, 147–148
  in environment, 211–213
  in human rights, 240–242, 241t
  in information technology, 167–172
  in libraries, 158
  in medicine, 50, 51t, 52
  against pests, 42–43
  in political freedom, 111–115,
    120–124, 229–230
  against racial discrimination, 9, 85,
    241
  rate of, 259, 264
  against sex discrimination, 85, 241
  in transportation, 185–186
  in women's rights, 241
  in working conditions, 69–73
prosperity
  and food, 31–38
  global trends in, 25–28
  in United States, 25
prostate-specific antigen (PSA), 62
protectionism, decline in, 140
PSA (prostate-specific antigen), 62
pumpkins, giant, 39, 40f

Q
Qatar, student-teacher ratios in, 152
QUASAR, 149, 150f

R
racial discrimination, advances against,
  9, 85, 241
radio-frequency identification, 102–103
radioisotope scans, 61
radio technology
  and air transport, 188
  and education, 159
railroads, 186
rain forests, 218
Ramos, Fidel, 119

Rand Corporation, 170
RCA, 104
Reagan administration
  and distribution of wealth, 23–24
  job growth during, 76
reaping, 38
recycling, 219
refrigerators, 94
Reich, Robert, 24
religious institutions, reaching out by,
  247–251
religious observance, 5, 9–10, 243–258
  and capitalism, 255–258
  future of, 265
  global trends in, 244–247, 245t
  and health, 10, 253–254
  and life enrichment, 251–255
  perspectives on, 244
  political opposition to, 245
  spread of, 243–247
renewable energy sources, 94–95, 95f,
  225
research and development, spending on,
  89
respect for humanity, progress in,
  240–242
retirement
  life expectancy and, 195
  trends in, 195–196
Reuters, 168
rice, production and cost, 40–41, 41f
river blindness, 4, 56
Rockefeller Foundation, 56
Rockwell International, 232–233
Rollins, John, 251
Rondeau, Glenn L., 71
Roosevelt, Franklin D., 52
Ruskin, John, 69
Russia
  capitalism in, 129
  defense industry conversion in,
    233–234
  democracy in, 120
  democratic supports in, 122
Russo, Vincent A., 132–133

S
Saddleback Valley Community Church
  (Mission Viejo, CA), 250
Saint Paul's School (Concord, NH), 154

Saints Peter & Paul Catholic Church
  (Hoboken, NJ), 247
Salaam, Rashaan, 252
Salisbury, Dallas, 75
Salk, Jonas, 52
Saltzman, Barbara, 178
Salvarsan, 62
Santa Fe Forestal y Industria, 132
satellite technology
  and air transport, 188
  and communications, 173
  uses of, 91–93
Savignac, Antonio, 199–200
scanners, 105
Schneider, Thomas F., 97
Schuylkill Falls (Philadelphia, PA), 96
Scott, Willard, 49
screwworm fly, 42–43
Sears Tower, 99
secondary education
  enrollment in, 148
  in Japan, 152–153
Second Baptist Church of Houston, TX,
  248
self-development, technology and, 7
self-discipline, religion and, 252–253
self-employment, 78, 82
self-help publications, commercial,
  166
Sensar, 102
service jobs
  growth in, 71–72, 72t, 77–78
  quality of, 77
  technology and, 72–73
sex discrimination, advances against,
  85, 241
sexual harassment, 9
Shakespeare, William, 15
Shanghai Securities Exchange, 137
shareholders, in United States, 141,
  141t
shares traded, growth in, 142t
Shenzen Stock Exchange, 137
Sidhu, Sanjiv, 99–100
Simon, Julian L., 1, 12, 51, 231
Singapore, television in, 208
skyscrapers, 97–99
slavery, 126
Slovakia, property rights in, 125
small business, 6, 75, 250

smallpox, eradication of, 50–52
smart gun, 102
smart phones, 175
Smith, Adam, 32, 127
smoking, and cancer, 58
Social Security system, 29
software, 99–100
  educational, 155–157
Software Publishers Association, 155
Sojourner, 91
solar power, 94–95, 95*f*
Sony Corp., 205
South Africa
  democracy in, 230
  democratic supports in, 123–124
  voting rights in, 126
Southern California Edison solar power
    plant, 95, 95*f*
South Korea
  education in, 153, 162
  private investment in, 139
  television in, 208
space exploration, 190
  benefits of, 90–93
Spain, college-entrance ratio in, 152
Spencer, Herbert, 147
spending
  on books, 198
  on demilitarization, 238*t*, 239
  on education, 148, 149*t*, 235–239
  on entertainment, 202, 204*t*
  on health, 57, 235–239
  on leisure, 8
  military, reduction in, 229, 232, 233*t*
  on research and development, 89
  rise in, 17
  on sporting goods, 196–197
spirituality, 243–258
  and business success, 255–258
  future of, 265
  and health, 253–254
  and media, 258
The Spitfire Grill, 258
sporting event attendance, 197–198
  versus religious observance, 244
sporting goods, spending on, 196–197
sports, technology and, 107–109, 108*f*
state park systems, 201–202
State University of New York, 151

Statistical Abstract of the United
    States, 32, 34, 57, 59, 89, 94,
    152–153, 173, 234
stockbrokers, increase in, 143
strokes, drug therapy for, 63
student-teacher ratios, 152
sudden infant death syndrome, 66
SuperBike, 108*f*
surgery, 58–59
Sweden
  cellular telephones in, 175
  child survival in, 53
  and Internet, 181*f*
  paid vacation time in, 194
Sweet, Leonard I., 250
Switzerland
  food variety in, 37
  and Internet, 181*f*
  voting rights in, 126
syphilis, 62

T
Taiwan
  democracy in, 120
  education in, 162
  food production in, 46
  labor costs per output in, 77
  television in, 208
Taoism, in China, 246
tax collection, and democracy, 121–122
Taylor, Orley, 157
team concept, 82
technology, 87–109
  and agriculture, 32, 38–41
  and business, 99–100
  and education, 154–161, 156*t*
  future of, 260–263
  improvements in, 87–89
  perspectives on, 6–7
  and productivity, 69, 72
  telephonic, 174
  and workplace injury reduction, 70
technology shifts, 88–89
  in energy, 94
teenage parenthood, 255
telecommuting, 82–84, 160
teleconferencing, 159
telegraph, wireless, 169
telemedicine, 66, 91

telephone(s), 172–176
    cellular, 174–175
    cost of service, 173
    expanded functions of, 104, 175
    future of, 174
    prevalence of, 17–18, 172, 175t
    technology, 174
teletype machines, 169
television, 176–177, 202–204
    in Asia, 208–210
    cable, 176, 202, 210
    and education, 159
    in emerging nations, 28
    prevalence of, 17–18
telework centers, 84
Tennyson, Alfred, Lord, 229
Tertullian, 212
Tesla, Nikola, 90–91
tetanus, 54
Thailand
    education in, 159–161
    literacy rate in, 159
    malaria in, 56–57
    private investment in, 139
    prosperity in, 26
    refugees in, 237
    student-teacher ratios in, 152
    tourism in, 200
theological students, trends among,
    257–258
Third World. *See* emerging nations
3M Corporation, 220–221
3TC (Epivir), 63
tilapia, 43
Timken Steel Co., 100
Tiwari, Narayan, 162
toilets, low-flow, 217
Tomalia, Donald, 90
tomato, genetically engineered, 44
toothpaste, 63
tourism, 198–201
    in Asia, 208
    and rain forest preservation, 218
transportation, 4, 185–191
    future of, 189–191
    modes of, 187t
    privatization of, 134–135
    progress in, 185–186
    technology and, 106–107

Transportation Research Institute, Uni-
    versity of Michigan, 190
travel, 198–201
truck driving, technology and, 106–107
tuberculosis, 54
Turkey, literacy rates in, 27
typewriters, electric, 171

U
ultrasonography, 61
Unabomber, 6
unemployment, 73–74
UNICEF, 56
United Airlines, employee involvement
    in, 79
United Kingdom
    air quality in, 214
    child survival in, 53
    conveniences in, 28
    employee training in, 165
    food variety in, 35–37
    higher education in, 153
    labor costs per output in, 77
    medicine in, 57
    paid vacation time in, 194
    prosperity in, 25–26
    and telephones, 172
    university graduates in, 148
United Nations, 27, 133, 159
    Conference on Disarmament, 239
    Development Programme, 39, 56, 162,
        234, 238
    and education, 162
    Programme for Vaccine Development,
        54
    Transitional Authority, 237
    World Conference on Education for
        All, 159
    World Tourism Organization, 199
United States
    automobile registrations in, 188t
    businesses in, 141
    demographics in, 138
    energy consumption in, 93–95
    exports of, 139
    food variety in, 37
    and Internet, 180, 181f
    labor costs per output in, 76–77
    life expectancy in, 49

United States (continued)
  modes of transportation in, 187t
  paid vacation time in, 194
  physician availability in, 57
  religion in, 244
  service jobs in, 78
  shareholders in, 141, 141t
  student-teacher ratios in, 152
  voting rights in, 126, 241
United States Army, recruiting, educa-
    tion and, 165
United States-Canada Air Quality Ac-
    cord, 227
United States Congress Office of Tech-
    nology Assessment, 155
United States Department of Energy, 95
United States Department of Trans-
    portation, 107
United States Export-Import Bank, 133
United States Federal Communications
    Commission, 158
United States Forest Service, 201
United States Small Business Adminis-
    tration, 250
United Theological Seminary, 250
University of California at Los Angeles,
    162
University of Michigan Transportation
    Research Institute, 190
University of Pittsburgh Medical Cen-
    ter, 254
University of Washington, 151
unmarried parenthood, 255
Uruguay, free trade in, 140
Uzbekistan
  cellular telephones in, 175
  property rights in, 125

V
Vaupel, James, 67
VCRs (video cassette recorders), 21, 202
vending machines, 103–104
venture capitalists, 140
Vera, Arturo, 138
Verrazano Narrows Bridge (New York,
    NY), 96
video cassette recorders (VCRs), 202
video conferencing, 177
Vietnam, property rights in, 125

Vincent, James, 66
voice mail, 174
Volpe, Welty & Co., 155
voting rights, 126, 241

W
wage controls, 129
Wall Street Journal, 6, 25, 118, 176
Walt Disney Co., 248
war, 9. See also peace
Washburn International, 247
Washington, Teddy, 252
water quality, 215–217
water use, 217, 217f
waterways, 212
wealth, 21–25
  and class, 24–25
  distribution of, 23–24
  future of, 260–261
  growth in, 22t, 22–24
  spiritual versus material, 256–257
  transfers of, and baby boomers, 29
wheat
  improvements in, 39
  production and cost, 40–41, 41f
whooping cough, 54
Willow Creek Community Church
    (South Barrington, IL), 247,
    249–250
wind power, 94
women
  enrollment of, in emerging nations,
    161
  job security of, 74, 75f
  rights of, progress in, 241
  voting rights of, 126, 241
word processors, 171
work
  challenge and interest of, 78–82
  and leisure, 193–196
workers
  compensation of, 7
  competitiveness of, 76–77
  increase in, 75
  involvement in management, 79, 131
  training of, 165–166
  unpaid, and GDP, 17
working conditions, 8, 69–85
  challenge and interest and, 78–82

future of, 262
improvements in, 69–73, 82–85
workplace absenteeism, 58
    employee involvement and, 79
workplace injuries, 70
workplace, stability in, 73–78, 75f
work time, purchasing power of, 18t,
    18–19
work week, 193–194
work year, 194
World Bank, 132, 165
World Economic Development Congress,
    133
World Health Organization, 52, 56
    Expanded Program on Immunization,
    54
World Trade Center, 99
World Trade Organization, 174
worms, parasitic, 54–56
Wriston, Walter B., 136

X
X–33, 190–191
xerography, 171

Y
Yellow Freight Systems, 83
Yeltsin, Boris, 120, 239
Yosemite Valley, 201
Young, Edward D. III, 158
youth population
    in emerging nations, 28–29, 138
    and entertainment, 204–205
    religious institutions and, 247

Z
Zedillo, Ernesto, 123
Zehr, Nathan and Paula, 40f
zoological gardens, 206
Zyuganov, Gennadi, 120